Fault Finder’s Bible: UK Electrical Edition

ISBN: 978-1-7641674-1-3
Second Edition
Printed in UK

Disclaimer:
This book is not a replacement for formal electrical training or licensing. It is designed as a supplemental resource for qualified professionals and apprentices. Always follow your local electrical codes, safety regulations, and manufacturer instructions. The author and publisher are not liable for any damages resulting from the application of this material.

FAULT FINDERS BIBLE: UK ELECTRICAL EDITION

REVISION 2

About the Author

Jordan Butler has worked in the electrical industry for over seven years, with hands-on experience across domestic, commercial, and industrial environments. His interest in electrical systems started at a young age, often taking apart old equipment to understand how it worked, a curiosity that naturally led to a career as a qualified electrician.

Throughout his time in the trade, Jordan identified a consistent gap: while installation methods and electrical theory are well covered in training, fault-finding is often left to trial and error. Apprentices and experienced electricians alike are frequently expected to "just figure it out" on site, without clear, structured processes to guide safe and efficient diagnosis.

That gap led to the creation of **Fault Lab**.

Jordan developed *The Fault Finder's Bible: Electrical Edition* to provide the kind of practical, real-world fault-finding guidance he wished he'd had as an apprentice. First released in Australia and New Zealand, the book has been widely adopted by electricians, apprentices, wholesalers, and training organisations as a reliable, site-ready reference for fault diagnosis.

Following its success across Australia and New Zealand, the book has now been adapted for the UK electrical industry, aligning terminology, testing methods, and references with BS 7671:2018+A4:2026 (IET Wiring Regulations) and UK site practices. The goal is to help apprentices and electricians fault-find logically, work safely, and diagnose problems with confidence, not guesswork.

You can follow the **Fault Lab** journey at:
Website: faultlab.co
Instagram / Facebook: @FaultLab

Introduction to Fault-Finding

Welcome to Fault Finder's Bible: UK Electrical Edition, your essential guide to mastering electrical fault finding.

Whether you're an apprentice, a seasoned electrician, or a business investing in upskilling your team, this book is designed to sharpen your diagnostic skills, boost your confidence on the job, and help you fault-find efficiently, safely, and systematically across domestic, commercial, and industrial environments.

Inside these pages, you'll learn:

- How to read electrical diagrams and schematics
- What common electrical symbols mean, and how to interpret them
- The step-by-step processes to follow when diagnosing electrical faults
- The tools and testing equipment you'll need, and how to use them correctly
- Common fault scenarios across different settings, with clear, repeatable procedures for identifying and resolving issues

You'll also find visual flowcharts, real-world examples, and troubleshooting tips developed from field experience. This book isn't hard to read, it's written by an electrician, to reflect the challenges and expectations you face every day.
The ability to fault-find effectively is one of the skills that sets great electricians apart. It's a skill in high demand, especially in high-pressure environments like construction sites, commercial maintenance, and heavy industry. This book is your training manual, reference guide, and confidence booster all in one.

Let's get into it. Turn the page and start becoming the kind of electrician people call when things stop working.

Safety Disclaimer: Live Circuit Testing

This guide includes guidance on voltage testing and fault diagnosis that may require work on or near live electrical equipment.

Warning: Live testing should only be carried out where it is necessary for fault diagnosis, it is unreasonable for the equipment to be made dead, and suitable precautions have been taken in accordance with the Electricity at Work Regulations 1989.

Before conducting any live tests, ensure:
- You are qualified and competent to perform live work.
- Live testing is necessary to diagnose the fault, and no safer alternative exists.
- Appropriate PPE (arc-rated clothing, insulated gloves, safety glasses) is worn.
- Test equipment is rated for live testing (Category III or IV as per IEC 61010).
- A suitable safe system of work has been established, and all non-essential personnel are clear of the testing area.

Primary UK References:

- BS 7671:2018+A4:2026 (IET Wiring Regulations, 18th Edition)
- Electricity at Work Regulations 1989 (UK)
- HSG85 – Electricity at Work: Safe Working Practices

In addition to the national framework above, electricians must comply with any site-specific rules, duty holder procedures, and recognised industry guidance (such as construction, industrial, or facilities management policies), which may impose stricter controls than the minimum statutory requirements.

Definition of Live Testing

Live testing is the process of carrying out electrical measurements or fault diagnosis on equipment while it remains energised and hazardous voltages are present. Unlike dead testing, live testing involves working on or near electrical equipment where an electric shock or arc flash hazard may exist. Live testing may be necessary to:

- Confirm the presence of voltage.
- Measure current under normal operating conditions.
- Identify voltage drops, phase loss, or imbalances.
- Verify the operation of protective devices and control circuits.

Disclaimer

Warning and Limitation of Liability

This book is intended as an educational reference only. It is not a substitute for formal training, manufacturer instructions, site procedures, or applicable legislation.

The information in this publication has been prepared using United Kingdom electrical standards and industry guidance available at the time of publication. It should be read alongside current legislation and recognised standards, including BS 7671:2018+A4:2026, Requirements for Electrical Installations (IET Wiring Regulations) and the Electricity at Work Regulations 1989.

Nothing in this publication should be interpreted as encouraging unnecessary live working. Dead testing and safe isolation should always be used wherever reasonably practicable.

Where live testing is necessary for fault diagnosis, it should only be carried out by competent persons using suitable test equipment, appropriate PPE, and a safe system of work in accordance with applicable legislation and site procedures.

Changes in regulations and technology

The electrical industry, including legislation, standards, and technology, is subject to change. The author and publisher accept no liability for any loss or damage arising from the use of the information contained in this guide due to changes in legislation, technological advancements, or changes in industry best practices.

Personal responsibility

As a qualified and competent electrician, you are responsible for the work you do. You must use your own professional expertise and judgement to determine the appropriate approach for any task. You should not attempt any work unless you have the necessary knowledge, skills, and experience to do it competently. Undertaking work outside your areas of competence is risky and can lead to unsafe situations.

No replacement for consultation or training

This guide does not replace the need for hands-on training, supervision, or consultation with competent persons. In case of doubt regarding any procedure or regulation, a qualified electrician should stop work and seek further advice from a more experienced professional.

Third-party content

Any external links, references, or third-party content included in this guide are for convenience only. The publisher and author do not endorse or take responsibility for the content, accuracy, or reliability of any linked websites or external sources.

Failure to work safely on or near electrical equipment may result in serious injury or legal consequences. Always follow safe isolation procedures wherever reasonably practicable.

TABLE OF CONTENTS

01 Introduction to Fault Finding

The Basics 11

The Dangers of Electricity 14

Safety and PPE 16

Lockout/Tagout Procedures 21

The Fault Finding Process 23

02 Testing Equipment

Multimeters - Basics and Advanced Usage 43

Insulation Resistance Testers 50

Clamp Meters 53

RCD Testers 56

Fault Loop Impedance Testing 58

03 Domestic Faults

RCD/RCBO Faults 62

Loss of Power on Circuits 66

Protective Device Operating 70

Hot Water Cylinder Faults 74

04 Commercial Faults

Power Distribution Systems in Commercial Buildings 79

HVAC Systems: Diagnosing Faults 83

Emergency Lighting Faults 89

TABLE OF CONTENTS

05 Industrial Faults

Motor Control Circuits and Faults 94

Control Panels and PLC Systems 98

Testing Faulty Sensors 105

Troubleshooting Heavy Machinery 108

How to Test a Motor 112

06 Conclusion and Final Checks

Verifying Repairs and Ensuring Safe Operation 120

Maintenance and Preventative Measures 122

07 Glossary and Quick References

Glossary 124

References 128

Diagram Symbols 130

Safety Checklist 132

Fault Finding Checklist 135

Notes 139

1

INTRODUCTION TO FAULT FINDING

The Basics

The Basics of Electrical Fault Finding

Electrical fault finding is the process of diagnosing and rectifying faults within an electrical system. Whether working on domestic, commercial, or industrial systems, understanding the fundamentals of electricity and common faults will make troubleshooting more efficient and effective.

Understanding How Electricity Works

Electricity flows through a closed-loop system known as a circuit.
A circuit consists of:
A power source (such as a battery or AC mains supply)
Conductors (cables that carry current)
Loads (devices that consume electricity, like lights or appliances)
Switches, isolators and protective devices (such as circuit breakers, RCDs. RCBOs and fuses)

Electricity follows Ohm's Law, which states that Voltage (V) = Current (I) × Resistance (R). This relationship helps in diagnosing faults by measuring changes in voltage, current, or resistance. A simple way to remember this is by referring to the illustration below.

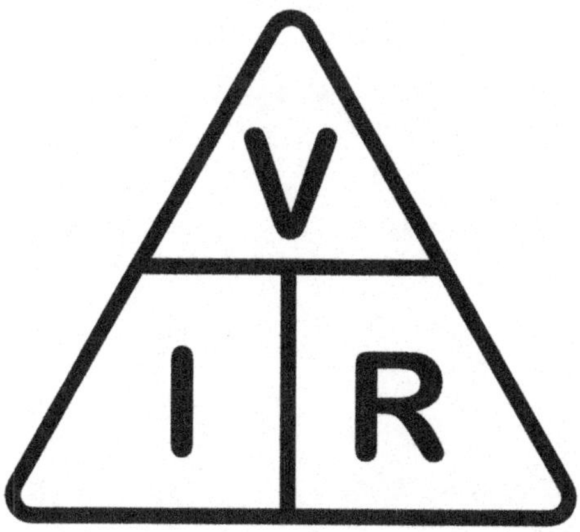

Common Electrical Faults and What to Look Out For

Several electrical faults can occur in any system, including:

- **Open Circuits** – A break in the circuit prevents current from flowing (e.g., a broken wire or faulty switch).
  - Signs: No power to the load, no continuity when tested.

- **Short Circuits** – A low-resistance path causes excessive current flow, often causing protective devices to operate.
  - Signs: Tripped circuit breakers, blown fuses, overheating.

- **Earth Faults** – An unintended path allows fault current to flow to earth.
  - Signs: RCD tripping, exposed conductive parts becoming live, or electric shock.

- **Overloads** – Too many devices drawing current on a circuit, causing protective devices to activate.
  - Signs: Circuit breakers tripping intermittently.

- **Voltage Drops** – Excessive impedance or resistance in the circuit wiring leads to reduced voltage at the load.
  - Signs: Dimming lights, underperforming motors.

AC vs. DC Voltage

Understanding the difference between Alternating Current (AC) and Direct Current (DC) is crucial for troubleshooting different systems:

- **AC** (Alternating Current)
  - Changes direction periodically (50Hz).
  - Used in mains electricity for homes and businesses.
  - Easier to transmit over long distances with transformers.
- **DC** (Direct Current)
  - Flows in one direction consistently.
  - Found in batteries, solar power, and electronic circuits.
  - Provides stable voltage, ideal for sensitive equipment.

Basics of How Transformers Work

Transformers are essential for stepping voltage up or down in AC electrical systems. They work on the principle of electromagnetic induction and consist of:

- **Primary winding:** Connected to the input voltage.
- **Secondary winding:** Provides the transformed voltage output.
- **Core:** Concentrates the magnetic field for efficient energy transfer.

Types of Transformers:

- Step-up Transformer: Increases voltage (e.g., high-voltage transmission networks).
- Step-down Transformer: Decreases voltage (e.g., consumer electrical installations).
- Isolation Transformer: Provides electrical separation for safety.

The Dangers of Electricity

Electricity is extremely useful, but also hazardous when not handled properly. Some of the main dangers include:

1. Electric Shock: When electrical current passes through the body, it can cause burns, muscle contractions, and even stop the heart.
2. Electrocution: A fatal electric shock.
3. Electrical Burns: Electrical current can cause severe internal and external burns.
4. Arc Flash & Arc Blast: A sudden release of energy caused by an electrical fault that can reach thousands of degrees, causing severe injuries.
5. Fires: Faulty wiring, overloaded circuits, and short circuits can ignite fires.
6. Explosions: Sparks in flammable environments (e.g., mines, petrol stations) can lead to explosions.

What Happens When an Electric Shock Occurs?

The effects of electric shock depend on current (amperes), voltage, duration, and pathway through the body. Possible effects include:

- 0.5 - 1mA: Slight tingling sensation.
- 1 - 5mA: Mild shock, can be painful but not dangerous.
- 5 - 10mA: Stronger shock, possible muscle contractions.
- 10 - 20mA: Loss of muscle control, difficulty letting go of live wires.
- 20 - 50mA: Pain, severe muscle contractions, difficulty breathing.
- *50 - 100mA: Ventricular fibrillation (irregular heart rhythm), can be fatal if untreated.*
- *100 - 200mA: Severe burns, heart stops beating effectively.*
- *200mA+: Heart can go into sustained contraction, leading to death.*

[Source: IEC 60479-1, Effects of Current on Human Beings and Livestock]
Actual effects vary depending on the current path through the body, duration of exposure, frequency, and individual physiological factors.

Lethal Current:

- Currents as low as 30 mA passing through the body may cause ventricular fibrillation under certain conditions.
- Currents around 100 mA present a very high risk of ventricular fibrillation and can be fatal.
- High-voltage shocks can cause deep internal burns and nerve damage, even if the current is momentary.

Why Does Current Kill, Not Just Voltage?

- Voltage (V) pushes the current (A) through the body, but it is current that causes harm.
- High voltage increases the risk of current flowing, but without sufficient current, serious injury may not occur.

Safe Working Practices

- Always isolate power before working on circuits.
- Wear proper PPE (insulated gloves, safety boots, arc flash protection if required).
- Use insulated tools and test for voltage before touching conductors.
- Never carry out live work unless it is justified, suitable precautions have been taken, and it is carried out in accordance with the Electricity at Work Regulations 1989 and site-specific procedures.
- Always comply with applicable legislation, recognised standards, and site-specific procedures.

Safety and PPE

Electrical work comes with various risks, from electric shocks and burns to falling objects and hazardous environments. Proper Personal Protective Equipment (PPE) and safety protocols help minimise these risks. Always comply with applicable legislation, site-specific requirements, and your employer's PPE policies.

1. Personal Protective Equipment (PPE) and Why It's Needed

Foot Protection – Safety footwear

- Why?
  - Protects feet from heavy objects (tools, materials, distribution boards) falling.
  - Reduces risk of injury from stepping on sharp objects like nails or screws.

Hand Protection – Insulated Gloves & General Work Gloves

- Why?
  - Insulated gloves (rated for voltage) reduce the risk of electric shock during justified live testing.
  - Leather or rubber gloves protect hands from cuts, abrasions, and burns.
  - Where insulated gloves are required, they should be selected and tested for the voltage of the work being undertaken.

Eye & Face Protection – Safety Glasses / Face Shield

- Why?
  - Protects eyes from sparks, arc flash, or flying debris when cutting conduit or drilling.

  - Reduces risk of dust or chemicals getting into eyes in industrial environments.
  - Arc-rated face shields are required when working on or near equipment where an arc flash hazard has been identified.

<u>Respiratory Protection – Dust Mask / Respirator</u>

- Why?
  - Prevents inhalation of dust, fibreglass particles, or toxic fumes when drilling, grinding, or working in confined spaces.
  - Respiratory protection should be selected in accordance with the task, site risk assessment, and applicable health and safety requirements.

<u>Hearing Protection – Earplugs / Earmuffs</u>

- Why?
  - Protects hearing in loud environments (e.g., working near generators, drilling into concrete).
  - Reduces the risk of long-term hearing damage in high-noise workplaces.

<u>Fire-Resistant (FR) Clothing & Arc-Rated PPE</u>

- Why?
  - Reduces the risk of burn injury from arc flash events.
  - Cotton or fire-retardant materials should be used instead of polyester (which melts onto the skin).
  - For areas with an arc flash risk, wear ATPV-rated (Arc Thermal Performance Value) arc flash suits.

<u>Head Protection – Hard Hat</u>

- Why?
  - Protects against falling objects (tools, materials, or overhead hazards).

Safe Isolation Equipment

- Why?
  - Ensures equipment is isolated and cannot be re-energised while work is in progress.
  - Prevents accidental energisation of circuits that could cause an electric shock.
  - Always use a personal lock and keep the key with you.

2. Safety Protocols for Electrical Work

Task-Specific Risk Assessment And Why It's Important

A task-specific risk assessment is a step-by-step assessment of risks before starting work. It helps identify:

- Potential electrical hazards.
- The right PPE and tools needed.
- Safe work procedures to follow.
- Emergency procedures in case of an accident.

A risk assessment should be completed before starting any new job, especially in high-risk areas.

Signing In & Out of Site

- Why?
  - Ensures everyone on-site is accounted for in case of an emergency.
  - Helps track who is working on specific electrical systems.
  - Often required on construction, industrial, manufacturing, and commercial sites.
  - Always follow the site's sign-in and sign-out procedures where required.

Electrical Isolation & Locking Off

- Why?
  - Prevents accidental re-energisation of circuits.
  - Dead working should always be the preferred method. Live work should only be carried out where justified and in accordance with the Electricity at Work Regulations 1989.

- Steps for Safe Isolation:
  - Identify the correct circuit.
  - Isolate the supply.
  - Secure the isolation (lock off where possible).
  - Attach a warning notice/tag.
  - Prove your voltage indicator on a known live source.
  - Test the circuit is dead.
  - Re-prove the voltage indicator on the known live source.

Note: Refer to correct isolation procedures on page 21.

Testing Before Touching

- Why?
  - Electrical circuits can remain live even when they appear off.
  - Some circuits have back feed from generators or UPS systems.
- Always test with a two pole voltage tester before touching any wires, also test on a known live source to confirm the supply is isolated.

Working at Heights Safety

- Why?
  - Falls are one of the leading causes of workplace injuries for electricians.
  - Ensure ladders are stable, secured, and rated for electrical work (fibreglass preferred).
- Never use aluminium ladders when working near live equipment.

Emergency Procedures

- Why?
  - Knowing what to do in case of electric shock, fire, or injury can save lives.
- If Someone Receives an Electric Shock:
  - Do not touch the person while they are in contact with electricity.
  - Turn off power immediately if possible.
  - Use a non-conductive object (wood, plastic) to move the person away if the power cannot be turned off.
  - Call emergency services (999 or 112).
  - If the casualty is unresponsive and not breathing normally, begin CPR and use an AED if one is available, following the instructions provided. Continue until the emergency services take over or the casualty shows signs of life.

Safe Isolation Procedures

Safe isolation is one of the most important safety procedures an electrician can carry out. Before beginning work on any electrical circuit or equipment, you must ensure it has been correctly isolated, secured against reconnection, and proved dead.

1. Identify all sources of the electrical supply

- Identify the circuit or equipment to be worked on.
- Confirm the correct point of isolation using drawings, labelling, or testing where necessary.
- Identify all possible sources of electrical energy, including alternative or back-fed supplies where applicable.

2. Isolate the Supply

- Switch off the correct isolator, switch, or protective device supplying the circuit.
- Ensure every source of electrical energy has been isolated before proceeding.

3. Secure the Isolation

- Apply a lock-off device where possible to prevent accidental re-energisation.
- Fit a warning tag stating that work is in progress.
- Retain the key to your personal lock at all times.

4. Prove Your Voltage Indicator

- Before testing the circuit, prove the voltage indicator on a known live source or proving unit to confirm it is operating.

5. Test the Circuit is Dead

Using an approved two-pole voltage indicator, test between:

- Line and Neutral
- Line and Earth
- Neutral and Earth

6. Re-Prove the Voltage Indicator

- Immediately after testing, prove the voltage indicator again on the same known live source or proving unit.
- This confirms the tester remained operational throughout the test.

7. Carry Out the Work

- Once the circuit has been safely isolated and proved dead, work may begin.
- If the scope of work changes or there is any doubt about the isolation, stop work and repeat the safe isolation procedure.

Restoring the Supply

Before re-energising the circuit:

1. Ensure all work has been completed.
2. Remove all tools and test equipment.
3. Confirm all covers, guards, barriers and protective devices have been refitted.
4. Ensure everyone is clear of the equipment.
5. Remove only your own lock-off device.
6. Notify anyone affected before restoring power.
7. Re-energise the circuit and verify correct operation.

Pro Tip

Never assume a circuit has a single source of supply. Always check for alternative or back-fed supplies, including generators, UPS systems, photovoltaic (PV) installations, battery energy storage systems (BESS), and automatic changeover arrangements. A circuit must never be considered safe until it has been correctly isolated and proved dead.

Reference: Electrical Safety First – Best Practice Guide 2: Safe Isolation for Low Voltage Installations.

The Fault-Finding Process

Initial Assessment – Identify the problem and gather necessary information.

Safety Checks – Ensure all safety protocols are followed before working on equipment.

Diagram Analysis – Study the wiring diagrams to understand how the circuit is designed to operate.

Testing – Use the right tools to test circuits and components.

Diagnosis – Isolate the fault by analysing test results and ruling out potential causes. Do not assume a fault based on a similar experience; still follow the correct procedure.

Repair – Rectify the issue and verify that the system is working as expected.

Final Check – Test the system under normal operating conditions to ensure everything is functioning correctly.

<u>Note</u>: **Please find an easy to use fault finding checklist at the rear of this book.**

Initial Assessment

When arriving at the fault, speak to the person who reported the fault and ask the following questions:

- What was happening when the fault occurred? e.g. there was a lightning strike nearby, or the operator could smell burning from the MCC room.
- Who was present when the fault occurred?
- Was anything turned on or off at the time leading up to the fault?
- What circuit is turned off? Or what piece/s of machinery are no longer operating correctly?

Review available documentation:

- Electrical drawings, wiring diagrams, and manuals
- Previous fault history and maintenance logs
- Manufacturer's troubleshooting guide

Identify the affected circuit or system:

- Is the fault in a single appliance, circuit, or an entire system?
- Is there any visible damage, such as burn marks, loose wires, or melted components?

Example: A customer reports that an RCD keeps tripping in a residential circuit. You need to determine whether the issue is caused by a faulty appliance, damaged wiring, or a faulty RCD/circuit breaker.

Safety Checks

Before attempting to diagnose and repair electrical faults, it is crucial to follow proper safety procedures to prevent injury, equipment damage, and breaches of safe working procedures. Electrical faults can pose serious hazards, including electric shock, arc flashes, and fire risks. The following safety checks must be completed before working on any electrical system:

1. Isolate the Power Source

Before touching any electrical equipment, always isolate the power supply. This can be done by switching off the relevant circuit breaker, disconnecting the power at the main switch, or removing fuses. Ensure that the power is completely cut off to prevent accidental shocks.

2. Verify the Absence of Voltage

Never assume a circuit is de-energised. Use an approved voltage tester or multimeter to confirm that no electrical current is present. Always prove the voltage indicator on a known live source or proving unit before and after testing the circuit.

3. Follow Safe Isolation Procedures

Safe isolation procedures are essential for preventing accidental re-energisation of equipment. Securely lock circuit breakers or switches in the "off" position and attach a tag indicating maintenance is in progress. Only the person who applied the lock-off device should remove it.

4. Use the Correct Personal Protective Equipment (PPE)
Wearing the right PPE can significantly reduce the risk of injuries. Assess what PPE is required for the task; if unsure, speak with a supervisor.

5. Check for Environmental Hazards
Before beginning any fault-finding task, assess the area for environmental conditions that could compromise safety. Wet or slippery surfaces significantly increase the risk of electric shock, particularly where electrical hazards are present. Ensure the work area is dry and stable, and use insulated mats or appropriate footwear when necessary.

Poor lighting, excessive heat, confined spaces, and overhead or underground services can also present serious risks. Always make sure the area is well lit and ventilated, and be cautious of any potential trip hazards, wasp nests, or vermin damage in switchboards, ceilings, or outdoor locations.

6. Correct Earthing
Ensure that the electrical system is correctly earthed to minimise the risk of fault currents or shocks. Earthing provides a safe path for excess electricity to flow in the event of a fault.

7. Task-Specific Risk Assessment and Site Sign-in
Before starting any work, complete a Risk Assessment or the required documentation to identify potential dangers and implement necessary control measures. Additionally, signing in and out of the worksite ensures that supervisors and emergency personnel are aware of your presence.

By following these safety checks, electricians can work confidently, safely and efficiently while minimising the risks associated with electrical fault-finding.

Diagram Analysis

Electrical diagrams are visual representations of electrical circuits or systems. These diagrams are crucial for electricians, engineers, and technicians to understand the design, function, and potential faults within a system. Mastering diagram analysis allows an electrician to efficiently troubleshoot faults, assess electrical systems, and ensure safe and reliable operation of electrical equipment. It is equally important to keep electrical drawings up to date whenever components or equipment are modified. Accurate drawings make future fault finding safer, faster, and help prevent costly mistakes.

Types of Electrical Diagrams

Schematic Diagrams

- Schematic diagrams use standardised symbols to represent electrical components and show the relationships between them. They are not concerned with physical layout but focus on the functionality and operation of the system. Components like relays, resistors, capacitors, and transformers are represented by specific symbols. Understanding these symbols is key to diagram analysis.

Single Line Diagrams (SLD)

- Single Line Diagrams simplify complex electrical systems by representing them with a single line, where each component is depicted symbolically. These diagrams are especially useful for industrial, commercial and distribution systems, showing how power is distributed across the system without showing every individual conductor of multiple lines for each connection.

Block Diagrams

- Block diagrams are used to show the main functional units of an electrical system. Each unit or component is represented as a "block," and connections between the blocks indicate the flow of electrical signals or power. Block diagrams are useful for high-level system analysis, particularly in control systems and larger systems with multiple interconnected parts.

Reading and Interpreting Diagrams

- Understanding Symbols and Legends
    - Electrical diagrams use symbols to represent components, such as resistors, capacitors, switches, and earthing. Familiarising yourself with the standard symbols is the first step in diagram analysis. Most diagrams will include a legend or key that helps you decode these symbols. Be sure to study this carefully to avoid mistakes in interpretation.
- Analysing the Wiring and Connections:
    - Understanding how wires connect components is essential for tracing current paths and identifying fault points. Pay attention to:
    - Connection points (nodes, junctions)
    - Wire numbers or labels
    - Wire colours (if noted – especially in site-specific diagrams)
    - Terminal designations (e.g., L1, L2, L3, N, PE, 21/22, A1/A2 on contactors)
    - Cross-referencing points (like wire markers that link to other parts of the diagram or panel)

Wiring diagrams show how components are connected with lines, which represent wires or conductors. These lines can indicate different types of connections:

- Solid Lines: Represent physical connections between components.
- Dotted or Dashed Lines: Indicate optional or secondary connections, or signal paths.
- Earthed Connections: Earthed components are often denoted by a special symbol or line that connects to the protective earth (PE).

Component Identification

Recognising what each component does is critical. For example, switches control the flow of current, while resistors limit current. Identifying components allows you to understand the purpose of each part in the circuit and how they interact with each other.

Power Flow Analysis

Power flow is determined by tracing the circuit from the supply through the control and load circuits. Some diagrams use arrows or directional symbols where required, but many rely on the arrangement of components and conductor identification.

Tracing the path of the supply through the circuit is crucial in understanding the behaviour of the system and troubleshooting faults. The flow of current through various components is typically shown with arrows, which can help identify areas of the system where power might be disrupted.

Cross-Referencing Multi-Page Diagrams

In larger systems, diagrams are often spread across multiple pages.

Look for:

- Page numbers or sheet references
- Cross-reference tags (e.g., "See Sheet 3, Line 14")
- Terminal block markers (often shown on both ends of a wire that passes between pages)
- Learning to trace a circuit or conductor across multiple pages is vital in complex troubleshooting.

<u>Fault-Finding Using Diagrams</u>

- Identifying Common Faults
- Faults often manifest as issues in power flow, such as:
    - Open Circuits: Breaks or disconnections in the wiring that prevent current from flowing.
    - Short Circuits: Unintended paths that allow current to flow where it shouldn't, often causing overloads or damage to components.
    - Overloaded Circuits: Diagrams can help identify the circuits and protective devices involved, making it easier to determine whether an overload condition may exist.

<u>Pro Tip:</u>

Keeping a record of faults with the dates and time can help with future problems and diagnostics.

Using Diagrams to Diagnose Faults

- To diagnose faults, you must compare the behaviour of the circuit in the diagram with how the system is functioning in reality:
  - Step 1: Start by reviewing the diagram to identify all power sources, connections, and load-bearing components.
  - Step 2: Look for any discrepancies between the expected behaviour (based on the diagram) and the observed behaviour (real-life test results).
  - Step 3: Trace the path of power from the source, looking for points where the circuit may be open or where the supply is no longer reaching the load.
  - Step 4: Focus on areas where components are likely to fail, such as switches, fuses, and breakers. With the circuit safely isolated and proven dead, check continuity using an appropriate continuity tester or low-resistance ohmmeter.

Physical vs. Functional Understanding

Many people new to diagram reading confuse the functional layout with the physical layout.
Remember that diagrams don't always show where things are physically located. A relay coil might be on one control panel, and the contact it switches could be on another. Understanding how things work together functionally is more important than where they are located physically, especially in schematic or single-line diagrams.

<u>**Note:**</u> Please find different symbols associated with diagrams at the rear of this book.

Testing

After completing your diagram analysis and forming a hypothesis about the fault, begin your testing phase. This step is critical, as it allows you to confirm suspicions, eliminate variables, and pinpoint the exact cause of failure. Always start with the least invasive testing methods and gradually move toward more direct testing if required. Safety remains a priority throughout.
Use the correct test equipment for the job, and make sure it's been checked for proper operation. Record your results if you're dealing with a complex fault, need to communicate findings with others, or may need to return to the site.

Start with the following checks and tests:

- **Safe Isolation and Verification:**
    - Ensure the circuit you're working on is properly isolated. Use an approved two-pole voltage indicator and proving unit to prove the circuit is dead. Don't rely solely on switches being off, test for dead before starting.
- **Continuity Testing (While circuit is isolated):**
    - Use an appropriate continuity tester to check conductors, fuses, and terminations. This helps identify open circuits, broken wires, or loose connections. For example, if a socket-outlet has no power, continuity testing can verify whether the phase conductor is intact from the consumer unit.
- **Insulation Resistance Testing:**
    - Test between line, neutral, and earth conductors to check for breakdowns or leakage. An insulation resistance value below the minimum acceptable value specified by BS 7671 may indicate damaged insulation, moisture ingress, or carbon tracking. **Always ensure sensitive equipment is disconnected before carrying out this test.**

- **Earth Fault Loop Impedance Testing:**
  - Measure the impedance of the fault path to ensure that if a fault occurs, the protective device will operate within the required disconnection time. High loop impedance can point to poor earth continuity, undersized cables, or corroded connections.

- **Voltage Testing:**
  - If safe to do so, measure the voltage at key points in the circuit. Start at the origin (e.g. consumer unit) and work toward the load. This helps you determine where the voltage stops, which often narrows down the fault location. Look for unexpected drops, phase loss, or incorrectly wired terminals.

- **RCD and Breaker Testing:**
  - Use an RCD tester to check the tripping times and current thresholds of residual current devices. If an RCD is nuisance tripping, this test will help determine if it's operating correctly or if there's a leakage fault downstream. For circuit breakers, check for mechanical wear, signs of overheating, or improper sizing.

- **Component Testing:**
  - Test individual components such as contactors, relays, thermostats, and sensors. Many faults come down to simple failures of these devices. Compare readings to manufacturer specifications where available.

Note: Inspection and testing should be carried out in accordance with BS 7671:2018+A4:2026 (Part 6) and recognised industry guidance.

- **Functional Testing (If Applicable):**
  - Where safe and possible, simulate system operation to see how the equipment responds. This can help uncover faults that only appear under certain loads or operational conditions.
  - When functional testing, ensure a suitable means of emergency stop or isolation is available, and that the person assisting understands how to operate it.

Diagnosis

Once testing has been completed, begin the process of diagnosing the fault by interpreting your test results and observations. This is where you shift from gathering information to making logical conclusions. Be methodical, don't jump to conclusions without evidence. Rushing this step is one of the most common causes of misdiagnosis.

Compare your test results to expected values, and consider what is normal for that system. Pay attention to patterns, such as voltage present at the supply side but not the load side, or low insulation readings between specific conductors.

Here's a step-by-step approach to guide your diagnosis:

Review Test Results Carefully:

- Go through each test you performed and ask: what does this result tell me? For example, if you measured voltage at the consumer unit but not at the appliance, where is it being lost? What components sit between those two points?

Narrow Down the Fault Area:

- Use a process of elimination to isolate the part of the circuit or system causing the issue. This might involve splitting the circuit, isolating loads, or testing in segments to localise the problem.

Cross-Check with the Diagram:

- Revisit the wiring diagram and compare it with what you've seen on site. Are there any components in the path you missed? Are the terminations and connections in the right place? Look for discrepancies between theory and reality.

Consider Common Fault Patterns:

Keep in mind the most common causes of faults:

- Loose or broken connections
- Water ingress or corrosion
- Mechanical damage to cables or components
- Overloaded circuits or incorrect cable sizing
- Failed components (e.g. contactors, sensors, fuses)

Rule Out One Thing at a Time:

- Avoid changing multiple things at once, this can confuse the diagnosis. Confirm each finding before moving on. If you're wrong, go back to the last known good reading or condition.

Think Logically, Not Emotionally:

- Don't let frustration, time pressure, or assumptions cloud your judgment. Stay objective, and always go back to what the test results are telling you.

Repair

Once the fault has been accurately diagnosed, the next step is to carry out the repair. This stage must be approached with care and attention to detail to ensure the fault is rectified safely, correctly, and in a way that prevents recurrence.

Always follow the applicable legislation and manufacturer guidelines when replacing or repairing components. Don't be tempted to take shortcuts, especially in systems where safety, compliance, or reliability is critical.

Here are key points to consider during the repair process:

Confirm Isolation Again:

- Before handling any cables or components, reconfirm that the circuit is isolated and proven dead. Apply a suitable lock-off device and warning notice where required as part of the safe isolation procedure.

Replace or Repair Faulty Components:

- If a component has failed (e.g. a relay, contactor, fuse, or switch), replace it with one of the same rating and specification. Avoid using oversized protection devices to "solve" a tripping issue, this is dangerous and non-compliant. If the repair involves a cable or conductor, use proper joining methods and insulation techniques.

Fix Root Cause, Not Just the Symptom:

- For example, if a fuse has blown, ask why it blew, was it due to overload, a short circuit, or a failing load? Simply replacing it without resolving the underlying issue may lead to repeated faults.

Ensure Correct Terminations and Tightness:

- Loose terminations are one of the most common causes of faults. After making any connections, use the correct torque settings if specified and double-check all terminations for security.

Document Any Changes:

- If your repair involves changes to the system layout, wiring, or control logic, make sure it's documented clearly. Update drawings if needed, and communicate the change with the relevant people (e.g. the maintenance team or site manager).

Test After Repair (Before Re-Energising):

- Carry out the appropriate inspection and testing before re-energising the circuit, in accordance with BS 7671 where applicable. This helps catch any installation faults or errors before the circuit is live again.

Final Check

After completing the repair, it's essential to perform a thorough final check. This ensures the system is functioning correctly, safely, and as expected under normal operating conditions. Never assume the job is finished just because the fault is gone, proper verification is what separates a rushed fix from a professional repair.

Go beyond a quick flick of the switch, test it properly, observe the system's full startup, and ensure all safety and control systems are operating as they should.

Include the following steps in your final check:

Re-Energise Safely:

- Inform others before re-energising.
- Restore the supply while observing for any abnormal signs such as sounds, smells, or unexpected behaviour. Be ready to isolate again if needed.

Functional Testing:

- Operate the equipment or circuit through its full range of normal functions. Check that all parts of the system are responding correctly, including control systems, sensors, and protective devices.

Monitor for Hidden Issues:

- Some faults may not appear immediately. Where possible, let the system run for a few minutes to ensure stability. Keep an eye on temperature, current draw, and any other signs of stress or overload.

Confirm the Original Fault is Resolved:

- If someone initially reported the fault, go back and explain what was found and fixed. Confirm that the system is now behaving correctly from their point of view as well.

Clean Up and Restore the Site:

- Tidy up any tools, offcuts, or packaging. Restore all covers, guards, and signage. Leave the work area as good as (or better than) you found it.

Log the Work Done:

- If required, write a fault report or service note detailing what was found, what testing was done, and what was repaired. This can be valuable for future maintenance or troubleshooting.
- Include the date, time, equipment repaired or replaced, test results, and any observations that may assist with future fault finding.

Real-World Example: Hydraulic Equipment Not Reversing

Fault Description:

While on-site, I was approached by the auto-electrician who reported that a piece of hydraulic equipment (fed from a 230V supply) was only moving forward, but would not operate in reverse when the reverse button was pressed on the hard-wired remote connected to the control panel.

Step-by-Step Walkthrough Using the Fault-Finding Process:

1. Initial Assessment

I began by speaking with the auto-electrician to get more details.
I asked:

What was happening when the fault occurred?
"We were operating the equipment when it just stopped going back. It still goes forward fine."

Who was there at the time?
"Just me. I was operating it myself."

Was anything recently changed or worked on?
"No changes, but it's been raining on and off."

What exactly isn't working?
"The reverse function from the remote, forward still works perfectly."

This narrowed it down to the control or signal side of the reverse operation.

2. Safety Checks

Before opening anything, I made sure the system was isolated where needed.

- I checked for stored energy in the hydraulic system.
- I verified the 230V control feed was isolated before working on the remote and control panel internals.

3. Diagram Analysis

I traced the wiring physically with the matching core numbers as there was no wiring diagram available. I then traced out a quick wiring diagram to match the control panel to help and for future use.

- Identified the wiring path from the remote push buttons to the control panel inputs.
- Confirmed that each direction (forward/reverse) had its own signal conductor back to the control panel.

4. Testing

I began testing the most likely components:

- Remote Push Button: Opened the remote housing and tested the contacts on the reverse button.
    - Contacts were working as expected, switching properly.
- Continuity Testing: Moved on to checking the wiring from the remote back to the control panel.
    - Found 2 of the 8 conductors had no continuity, including the one responsible for the reverse signal.
- Visual Inspection: The cable showed signs of wear, likely from repeated bending or physical damage.

5. Diagnosis

The fault was isolated to the damaged multi-core cable between the remote and the control panel.

- The reverse signal wire was one of the damaged conductors.
- No issues were found with the remote, control relays, or actuators, just the cable.

6. Repair

I replaced the damaged cable entirely (about a 10-metre run as I could not visually see where the damage occurred), using a cable of the same specifications and suitable strain relief to prevent repeat damage.

- All terminations were redone and tested.
- Verified continuity on all 8 cores before re-energising.
- Reinstalled remote and secured all enclosures.

7. Final Check

After re-energising:

- Tested both forward and reverse functions on the remote.
- Confirmed the system operated correctly under normal conditions.
- Observed operation over several cycles with no further issues.
- Reported the cause and fix to the auto-electrician and site supervisor.
- Cleaned the area, labelled the new cable, and noted the repair in the job log.

Key Lesson:

Even simple faults like a reverse function not working can lead to time-consuming downtime if not approached systematically. Starting with the basics and working step by step, without assumptions, helped isolate the fault efficiently.

2

TESTING EQUIPMENT

Multimeters

Basic Uses of a Multimeter

A multimeter is an essential tool for electricians and technicians to measure electrical values. It typically has four main functions:

1. **Measuring Voltage (V)**
   - DC Voltage (⎓V): Used for batteries, power supplies, and electronics.
   - AC Voltage (~V): Used for socket-outlets, mains power, and appliances.
   - How to Use: Set to the correct voltage range, place probes on the circuit, and read the display.
2. **Measuring Current (A)**
   - DC Amps (⎓A): Used for circuits with batteries or DC power sources.
   - AC Amps (~A): Used in appliances, motors, and AC systems.
   - How to Use: Set to the correct range, connect in series with the circuit (A clamp meter is used without breaking the circuit.).
3. **Measuring Resistance (Ω)**
   - Measures the resistance of components or conductors. Useful for checking resistors, windings, heating elements, and other electrical components.
   - How to Use: Set to Ω, touch probes across the component, and check resistance value.
4. **Continuity Testing (Beep Mode)**
   - Use an appropriate continuity tester or low-resistance ohmmeter on isolated circuits to verify a continuous electrical path.
   - How to Use: A low resistance or audible tone indicates continuity. A high or infinite resistance indicates an open circuit.

Advanced Uses of a Multimeter

Beyond basic voltage, current, and resistance measurements, a multimeter can be used for advanced troubleshooting and diagnostics in electrical and electronic systems. The following sections explore these advanced features and their practical applications.

1. Measuring Frequency (Hz)
Purpose: Checking the frequency of AC power sources, motor drives, VSDs and generators.
How to Use:

- Set the multimeter to Hz mode.
- Place the probes across the AC supply or signal source.
- Read the frequency value on the display.

Example: Ensuring a generator output is at the correct 50 Hz.

2. Measuring Capacitance (F - Farad)
Purpose: Diagnosing capacitors in power supplies, motor start circuits, and electronic devices.
How to Use:

- Set the multimeter to capacitance mode (F).
- Verify the capacitor has been safely discharged before connecting the test leads.
- Connect the probes to the capacitor terminals.
- Read the capacitance value and compare it to the rated value.

Example: Checking a motor run capacitor in an air conditioning unit to verify that its capacitance is within the manufacturer's specified tolerance.

3. Measuring Duty Cycle (%)

Purpose: Diagnosing PWM (Pulse Width Modulation) circuits, motor controllers, and dimmers.

How to Use:

- Set the meter to duty cycle mode (%).
- Probe the PWM signal source.
- Read the percentage of the "on" time versus "off" time.

Example: Testing the duty cycle of a variable-speed drive (VSD) to ensure it is operating correctly.

4. Measuring Temperature (°C/°F)

Purpose: Monitoring heat buildup in electrical components, checking HVAC temperatures, or diagnosing overheating circuits.

How to Use:

- Connect a thermocouple probe to the multimeter.
- Set the meter to temperature mode.
- Place the thermocouple on the component to measure.
- Read the temperature on the display.

Example: Checking the temperature of a distribution board to diagnose potential overheating issues.

5. Measuring Low-Voltage Millivolts (mV)

Purpose: Diagnosing small voltage signals in sensitive circuits, including sensors and control systems.

How to Use:

- Set the multimeter to millivolt (mV) mode.
- Place the probes across the small voltage source.
- Read the display.

Example: Measuring the output of a thermocouple sensor in an industrial process to check for correct operation.

6. Testing Inductance (L) (Advanced Models)

Purpose: Diagnosing coils, transformers, and inductors for faults.

How to Use:

- Use a multimeter with inductance measurement capabilities.
- Set to inductance mode (L).
- Connect probes to the inductor or transformer winding.
- Read the inductance value.

Example: Checking the inductance of a coil or transformer winding to identify shorted turns or winding faults.

7. Non-Contact Voltage Detection (NCV Mode)

Purpose: Quickly identifying live AC voltage without direct contact for safety checks.

How to Use:

- Activate NCV mode on the multimeter (if available).
- Hold the meter near a suspected live wire.
- The meter will beep or flash if voltage is detected.

Example: Using an NCV tester to identify the possible presence of AC voltage. An NCV tester must never be used to prove a circuit is dead.

8. Diagnosing Harmonics in Power Systems (Advanced Models)

Purpose: Analysing power quality and detecting harmonics that can cause overheating and interference in electrical systems.

How to Use:

- Use a multimeter or power quality analyser with harmonic analysis features.
- Measure voltage or current distortions.

Example: Diagnosing harmonic issues in a variable speed drive (VSD) that is causing unexpected motor heating.

9. Measuring DC Offset in Power Supplies

Purpose: Identifying unwanted DC voltage in AC circuits that can cause interference and equipment malfunction.

How to Use:

- Set the meter to DC voltage mode.
- Place probes across the AC circuit output.
- If DC voltage is present, troubleshoot further.

Example: Checking an audio amplifier for unwanted DC offset that can cause speaker damage.

When choosing a multimeter, it's important to consider quality, durability, and local availability. Below are some of the most reputable multimeter brands used by professionals and electricians in the UK.

Fluke

- Overview: Fluke is the industry standard for electricians, known for its accuracy, rugged design, and safety features.
- Best For: Professional electricians, industrial applications, and precision testing.
- Popular Models: Fluke 87V, Fluke 117, Fluke 325 (clamp meter).

Megger

- Overview: Megger is a highly respected UK-based brand, best known for electrical testing and inspection equipment, with a strong focus on safety, reliability, and compliance.
- Best For: Inspection and testing electricians, EICRs, commissioning work, and fault-finding where insulation and earth testing are critical.
- Popular Models: MFT1800 series, MIT300 series.

What is a CAT Rating?

The Category (CAT) rating is a safety classification that defines where a multimeter/instrument can be safely used and how much transient voltage (spikes or surges) it can handle. These ratings are crucial for electrical safety, ensuring that the meter is designed for the type of environment you're working in.

Multimeters are classified into four CAT (Category) ratings:

CAT I – Low Energy Circuits

- Description: Used for electronic devices and low-energy circuits, such as electronic equipment that is not directly connected to the mains supply.
- Examples: Small electronics, circuit boards, and signal testing.
- Voltage Ratings: Various voltage ratings but only safe in low-energy environments.
- **Risk**: No protection against power surges from mains electricity.

CAT II – Standard Household Appliances

- Description: Designed for testing electrical appliances that are connected into socket-outlets.
- Examples: Power tools, extension cords, office equipment.
- Voltage Ratings: Varies up to 1000V (depending on the instrument).
- **Risk**: Limited protection from power surges. Not intended for measurements on fixed installation wiring or distribution equipment.

CAT III – Industrial and Commercial Applications

- Description: Suitable for measuring directly at the distribution board or final circuits.
- Examples: Distribution boards, motor control centres (MCCs), three-phase motors, and industrial or commercial installations.
- Voltage Ratings: 600V or 1000V (depending on the instrument).
- Risk: Protects against high transient voltages that occur in industrial environments.

CAT IV – High Energy, Utility-Level Testing

- Description: Used for main power lines, electricity meters, and outdoor installations at the origin of the electrical installation.
- Examples: Transformer connections, underground cables, and service heads.
- Voltage Ratings: 600V to 1000V or higher.
- Risk: Designed to withstand large power surges and protect against the highest electrical hazards.

Insulation Resistance Testers

What is an Insulation Resistance Tester?

An insulation resistance tester, commonly called a Megger, is a specialised device used to measure the resistance of electrical insulation. It applies a high DC voltage (typically 250V, 500V, 1000V or more) between conductors and between conductors and earth to test how well insulation resists leakage current. This test helps identify deterioration, moisture ingress, contamination, and breakdown in electrical systems before failures occur.

Refer to BS 7671:2018+A4:2026, Part 6, for requirements relating to insulation resistance testing and applicable test voltages.

Common Uses of an Insulation Resistance Tester

1. Electrical Wiring and Cable Testing
    - Ensures insulation integrity in mains wiring, underground cables, and distribution boards.
    - Identifies damaged or degraded wiring before failure occurs.
2. Motor and Transformer Testing
    - Detects insulation breakdown in motor windings and transformer coils.
    - Helps diagnose short circuits and insulation deterioration.
3. Testing Distribution Boards and Switchgear
    - Evaluates insulation in circuit breakers, busbars, and control panels.
    - Helps verify the condition of insulation in accordance with BS 7671 where applicable.

<u>**Note**</u>: Insulation resistance testing must only be performed on a **de-energised** circuit. Always confirm zero voltage, prove isolation, and ensure the circuit is safe before testing.

4. Generator and UPS System Maintenance
    - Prevents insulation failure in backup power systems.
    - Reduces risk of power disruptions.
5. Industrial and Commercial Applications
    - Essential for hazardous environments where insulation failures can cause serious accidents.
    - Used in HVAC industrial process plant.

How to Use an Insulation Resistance Tester

1. Safety Precautions

- Always disconnect the circuit from the supply and verify it has been isolated and proved dead before insulation resistance testing.
- Discharge all capacitors and stored charges in the circuit.
- Warn others that insulation testing is about to take place.
- Never touch the test leads or circuit under test during measurement.

2. Set Up the Tester

- Select the appropriate test voltage for the equipment or circuit:
    - 250V for extra-low voltage circuits (SELV, PELV)
    - 500V for low-voltage single-phase and three-phase circuits
    - 1,000V where required by BS 7671 or the equipment manufacturer.

3. Connect the Test Leads

- Connect one lead to the conductor you wish to test.
- Connect the other lead to either:
    - Earth, or
    - Another conductor (when testing between conductors).

4. Perform the Test

- Press the test button to apply the test voltage.
- Observe the resistance reading.
- A healthy circuit should typically show a resistance much higher than 1 MΩ, refer to BS 7671:2018+A4:2026, Part 6, for requirements relating to insulation resistance testing and applicable test voltages.

5. Interpret the Results

- High resistance (e.g. >1 MΩ): Good insulation.
- Low resistance (e.g. <1 MΩ): Possible deterioration, moisture ingress, or damage. Do not energise. Inspect further.
- Very low resistance: Short circuit or severe insulation failure. Do not energise. Inspect further.

If the resistance is lower than acceptable, investigate further and repair before energising the circuit.

6. After Testing

- After testing, safely discharge any stored energy in accordance with the manufacturer's instructions before reconnecting equipment.
- Remove test leads.
- Record the test results where required for maintenance records or inspection and testing documentation.

Pro Tip:

Before using an insulation resistance tester, consider whether the circuit contains any sensitive electronic equipment or is connected to an RCD. Sensitive devices can be damaged by the high test voltage, and RCDs may cause unexpected low or failed readings. Isolate or disconnect these components before testing whenever possible.

Clamp Meter

A clamp meter is a versatile diagnostic tool used for measuring electrical parameters without the need for direct contact with conductors. Unlike traditional multimeters, clamp meters allow electricians to take current measurements safely and efficiently, making them an essential tool for fault-finding in various electrical systems.

How a Clamp Meter Works
A clamp meter operates using a Hall Effect sensor or a current transformer to detect magnetic fields around a conductor and convert them into an electrical reading. This allows for non-contact current measurements, eliminating the need to break the circuit.

Uses of a Clamp Meter in Fault Finding
Measuring AC and DC Current

- Used to check if a circuit is overloaded by measuring the current draw of equipment.
- Helps determine whether the protective device is correctly rated for the load.
- Essential for diagnosing unbalanced loads in three-phase systems.

Detecting Faulty Components

- A sudden drop in current may indicate a failed motor winding or an open circuit.
- An excessive current draw may indicate an overloaded circuit, a mechanical fault, a developing motor fault, or another abnormal operating condition requiring further investigation.

Checking Inrush Current

- Measures the initial surge of current when a motor or transformer starts.
- Useful for diagnosing tripping issues with protective devices.

Identifying Leakage Current

- Measures small leakage currents flowing to earth, helping to detect insulation faults before an RCD trips.
- Helps diagnose nuisance tripping of RCDs.
- Requires a dedicated leakage current tester or clamp meter designed to accurately measure very low current levels.

Measuring Voltage and Resistance (Advanced Models)

- Some clamp meters have multimeter functions, allowing for voltage, resistance, and continuity checks.
- Useful when troubleshooting power supply issues alongside current measurements.

Best Practices for Using a Clamp Meter

✔ Ensure the meter is rated for the voltage and current being measured.

✔ Only clamp around a single live conductor. Clamping around both the line and neutral conductors together will normally result in a near-zero reading because the magnetic fields cancel each other.

✔ Position the clamp properly, placing it near high-interference areas can affect accuracy.

✔ Use the correct measurement mode (AC or DC) based on the application.

✔ Verify readings with a secondary test if values seem inconsistent.

Common faults to detect with a clamp meter

Overloaded Circuit	Measure the load current and compare it with the rating of the protective device and the cable
Transformer Faults	Compare primary and secondary currents or check for current imbalance between phases.
Motor Failure	Check for high inrush or low current draw
Nuisance RCD Tripping	Measure leakage current to earth
High Resistance Connections	Compare voltage drop and current between connections

RCD Testers

Residual Current Devices (RCDs) are critical safety components designed to protect against electric shock and prevent electrical fires. An RCD Tester is used to verify the correct operation of RCDs by measuring their trip times and sensitivity.
Understanding how to use an RCD tester is essential for electricians diagnosing faults in domestic, commercial, and industrial electrical systems.

How an RCD Works
An RCD continuously monitors the balance between line and neutral currents in a circuit. If an imbalance occurs (typically caused by current leakage to earth), the RCD disconnects power within milliseconds to prevent harm.

- Typical Ratings: 10mA, 30mA, 100mA, 300mA, and 500mA.
- Standard Trip Times:
    - 30mA RCD: Must trip within 300ms at rated current.
    - 30mA RCD: Must trip within 40ms at 5x rated current (150mA).

Pro Tip: When carrying out RCD trip-time (injection) tests, testing should be performed at an appropriate point on the final circuit, such as the furthest socket-outlet, to verify that disconnection times are achieved under fault conditions at the the furthest point of the final circuit.
Testing at the distribution board may be used to confirm correct operation of the RCD device, but it does not demonstrate protection of the final circuit wiring or accessories.

The integral test button on the RCD provides a functional check of the mechanism only and must not be relied upon as a substitute for instrument testing.

Uses of an RCD Tester in Fault Finding

Verifying RCD Operation

- Ensures the RCD trips within the required time limits.
- Confirms that the device is functional and not stuck or sluggish.

Checking Nuisance Tripping

- Identifies whether an RCD is overly sensitive or faulty.
- Confirms whether the RCD is operating correctly. If the RCD passes testing, further investigation should be carried out for earth leakage or faulty connected equipment.

Testing Earth Fault Protection

- Confirms that the RCD correctly disconnects power in an earth fault situation.
- Confirms that the RCD operates within the disconnection times specified by BS 7671.

Measuring Trip Time & Sensitivity

- Checks the RCD's response time under different conditions (1/2x, 1x, 5x rated current).
- Ensures it meets the required disconnection time.

Best Practices for Using an RCD Tester

– Test at the origin of the circuit first before testing downstream.
– Use the correct test setting (A, B or F type RCDs).
– Perform tests at both 0° and 180° phases for accurate results.
– Record results to compare against regulatory limits.
– If an RCD fails, inspect wiring and connected loads for leakage issues.

Note: Type AC RCDs are no longer recommended for many modern installations because they may not detect pulsating DC residual currents produced by electronic equipment. Type A RCDs are generally preferred for new installations and when replacing existing devices, in accordance with current UK practice and BS 7671.

Fault Loop Impedance Testing

What is Fault Loop Impedance (Zs)?

Fault loop impedance refers to the total impedance of the fault path in an electrical system, from the line conductor, through the fault, along the circuit protective conductor (CPC), and back to the supply transformer via the earthing system.

When a fault occurs (such as a line conductor contacting a metallic enclosure), sufficient fault current must flow to operate the protective device (circuit breaker or fuse) within the required disconnection time. If the impedance of the fault loop is too high, the resulting fault current may be insufficient to operate the protective device, leaving exposed conductive parts at a dangerous potential.

This is why fault loop impedance testing is critical for safety and forms part of the inspection and testing requirements of BS 7671:2018+A4:2026 (Part 6). particularly when commissioning new installations or verifying compliance following alterations or repairs.

Key Terms

- Zs = Total earth fault loop impedance (measured at the point of utilisation or distribution board).
- Ze = External earth fault loop impedance (measured at the origin of the installation).
- R1 + R2 = Resistance of the line conductor and circuit protective conductor (CPC).
- Ia = The fault current required to operate the protective device within the specified disconnection time.

How to Test Fault Loop Impedance

Tools needed: Loop impedance tester or multifunction tester rated for the circuit.

Step-by-Step Process:

1. Isolate the circuit if required for setup. Confirm it's safe to test live and justifiable.
2. Insert test probes or plug the tester into the socket-outlet or the point being tested.
3. Select "Loop Impedance" mode on the tester (often marked as “Zs”).
4. Perform the test, the tester injects a test current and measures the resulting voltage drop to calculate Zs.
5. Record the reading and compare it against the maximum allowable Zs value for the protective device used (see Table 41.2, 41.3 in BS 7671:2018+A4:2026).
6. Repeat testing on other points if needed (e.g., furthest point on the final circuits.

What’s a Pass? What’s a Fail?

- If Zs is too high, the fault current won’t reach the required level to operate the protective device fast enough.
- You’ll need to investigate:
  - Long cable runs (voltage drop)
  - Undersized conductors
  - Loose or corroded terminations
  - Poor earth continuity

Pro Tip: A fault loop impedance test doesn’t just check numbers, it confirms the earthing system is intact. If the circuit protective conductor, main earthing conductor, or supply earth connection is compromised, fault current may not return effectively to the source, preventing protective devices from operating within the required disconnection times.

Example:

You're testing a final sub-circuit protected by a 20A Type C RCBO. The required disconnection time is 0.4 seconds, and the maximum permissible Zs for that breaker is 1.09Ω.

The tester reads 1.8Ω. That's too high.

<u>Solution</u>

- Check all terminations first. If the high loop impedance reading persists, isolate the circuit and split the circuit at an accessible point.
- Test continuity on each section, this will help identify which side of the circuit contains the high-resistance fault, you can then hunt down to find which part of the circuit is giving the fault.
- Confirm the conductor size is appropriate for the circuit design and protective device.
- Check for long cable runs or poor connections that may be increasing the earth fault loop impedance.
- Measure the external earth fault loop impedance (Ze) at the origin of the installation. If Ze is higher than expected, the issue may lie with the supply or earthing arrangement rather than the final circuit.

3

DOMESTIC FAULTS

RCD/RCBO Faults

Residual Current Devices (RCDs) and Residual Current Breakers with Overload (RCBOs) are essential for electrical safety, but they can sometimes operate unexpectedly or fail to operate correctly. Understanding the common causes of RCD/RCBO faults and how to diagnose them effectively is crucial for electricians troubleshooting domestic, commercial, and industrial systems.

Step-by-Step RCD/RCBO Fault Diagnosis

Visual Inspection

- Check for visible damage, moisture, or burns around the RCD/RCBO.
- Inspect wiring for loose or damaged connections.
- Look for signs of water ingress, particularly in outdoor installations.

Checking for Load-Related Issues

Turn off all circuits protected by the RCD and attempt to reset the device with no loads connected.

If the RCD resets successfully with all loads disconnected, the fault is likely present on a downstream circuit or within a connected appliance. Plugged-in appliances are one of the most common sources of earth leakage, particularly those with internal heating elements, moisture ingress, or degraded insulation (e.g. kettles, washing machines, or refrigerators).

It's essential to unplug all appliances from affected circuits during the testing process, simply turning them off at the switch is not sufficient, as many still allow some leakage current to flow. Once unplugged, turn each circuit back on one at a time, and monitor the RCD response. If the RCD trips when a specific circuit is energised (especially with appliances connected), the issue is likely within that load path.

Failing to isolate these appliances properly can lead to misdiagnosis and wasted time, always remove all loads from the circuit before assuming the fault lies in wiring or the RCD itself. If the RCD still fails to reset with all circuits and loads disconnected, it may indicate a fault in the RCD itself, damaged cables due to vermin or incorrect wiring at the consumer unit.

Testing for Neutral-to-Earth Faults

- Isolate power to circuit
- Disconnect the neutral wire from the RCD (ensure correct isolations are in place first).
- Use an insulation resistance tester (megger) between neutral and earth.
    - A low resistance reading indicates a fault.
- Check for incorrect neutral wiring, especially in consumer units.

Measuring Earth Leakage Current

- Use a leakage current clamp meter capable of measuring milliamps (mA).
- If leakage is high, isolate circuits and test appliances individually.
- Leakage current is often kept below approximately 30% of the RCD's rated residual operating current to reduce the likelihood of unwanted tripping. This is a practical recommendation rather than a requirement of BS 7671.

Testing the RCD/RCBO with an RCD Tester

- Perform 1/2x, 1x, and 5x trip tests to check response times.
- If the RCD or RCBO fails the instrument tests, first confirm the correct test settings have been selected and that the device is wired correctly. If it still fails, replace the device and repeat the tests.

Neutrals Swapped on RCDs

One of the easiest faults to miss is neutrals connected to the wrong RCD. When this happens, the RCD sees an imbalance (current going out on the line conductor but coming back on a different neutral) and trips as soon as a load is connected.

Key Takeaways

✔ Always check for neutral-to-earth faults before replacing an RCD/RCBO.

✔ High earth leakage is a common cause of nuisance tripping, measure it correctly.

✔ If an RCD fails a test, confirm correct wiring before assuming it is faulty.

✔ Water damage and worn insulation are frequent culprits in outdoor and industrial installations.

✔ Disconnect and reconnect final circuits one at a time to quickly identify which circuit is causing the RCD or RCBO to operate.

Common Faults Diagnosed with an RCD Tester

RCD/RCBO not operating	Perform full-current (1x) test to confirm failure
Unwanted tripping	Use ramp test to check trip sensitivity
Nuisance tripping on circuits with motors	Measure leakage current and check for harmonics
Delayed or slow operation	Measure at ½, 1, and 5 times and compare to expected results in BS7671
No earth fault protection	Test RCD on multiple socket-outlets to confirm disconnection

Loss of Power on Circuits

When a circuit isn't delivering power, it can be frustrating, but by following a systematic approach, you can quickly pinpoint the issue. Here's a step-by-step guide to help you diagnose a "Loss of Power" situation:

Before starting the diagnosis, gather key details from the homeowner or client. Their insights can help you identify the cause of the "Loss of Power" issue faster:

- Recent Work: Ask if any electrical work, like installations or repairs, was recently done, or if new appliances were added.
- Power Surges or Outages: Check if there were recent power surges, storms, or local outages.
- Circuit Interruptions: Find out if the power loss was gradual or sudden, and if it affected one area or the entire property.
- Specific Appliances: Ask which appliances or socket-outlets are not working, or what appliances were being used before the power loss to help pinpoint the issue.

1. Ensure the Basics:

- Power Supply Check: Before you dive into the details, make sure there is power coming into the system. Check the supply is present at the origin of the installation to confirm the circuit is supposed to have power. If there's an issue at the distribution board, it could affect multiple circuits.
- Circuit Breaker/RCBO: Ensure that the protective device hasn't operated. A circuit protective device could be the result of a short circuit, overload, or a fault in the circuit. If a protective device has operated, consider the possible cause before resetting it. If it operates again immediately, investigate the fault rather than repeatedly resetting the device.

2. Check the Switches:

- Switch Operation: Make sure that the switch controlling the circuit is actually in the "on" position. It's easy to overlook a switch or a dimmer switch, which can cause no power on the circuit.

3. Test the Socket-Outlet:

- Voltage Testing: Use a multimeter to test the voltage at the socket-outlet. If you're getting no reading, it's a strong indicator that the problem is somewhere upstream in the circuit. If you're getting a reading, but it's low, this could indicate a partial fault (e.g., a loose connection).
- Loose Connections: Loose or faulty connections in the socket-outlet can lead to no power. Check for any burnt or loose wires inside the socket-outlet, and tighten as necessary.

4. Test the Circuit Wiring:

- Continuity Test: Isolate the circuit being investigated. Use a continuity tester, check the wiring from the socket-outlet back to the distribution board. This will help you locate any breaks or open circuits in the wiring.
- Inspect the Junction Boxes: If the wiring runs through junction boxes, check these areas for loose or disconnected wires. Junctions are common spots for issues to develop, such as corroded or damaged connectors.
- With the circuit safely isolated and proved dead, temporarily link the conductors together at one end of the circuit (or use a trailing lead where appropriate). Then move to the opposite end and carry out a continuity test. If continuity is not present, progressively divide the circuit into smaller sections until the fault is located.

5. Check for Overloaded Circuits:

- If the circuit breaker/RCBO is tripping, it could be due to an overload condition. Reduce the load on the circuit by disconnecting loads and restoring the protective device. If the protective device stays on, you may have identified the issue as an overloaded circuit.

6. Test the Circuit Breaker/RCBO:

- Protective Device Fault: If the protective device continues to operate without any apparent fault on the circuit, inspect it for signs of damage, overheating, or loose terminations. Confirm the wiring is correct and, where necessary, replace the protective device with one of the same type and rating before re-testing the circuit.

7. Inspect for Earth Faults:

- RCDs: If the circuit is protected by an RCD (Residual Current Device), ensure that it has not tripped. RCDs are designed to cut power if there's an imbalance between the line and neutral conductors, typically due to an earth fault.

- Leakage or Short Circuit: Carry out insulation resistance testing and inspect the circuit for earth faults, damaged insulation, moisture ingress, or incorrect neutral-to-earth connections. If the insulation on any wire has broken down, it could cause leakage, tripping the RCD.

8. Look for Visible Damage:

- Inspect the wiring for any visible signs of damage, such as burns, fraying, or chewed insulation (especially in places where rodents might be present). Damaged conductors or cables are a common cause of power loss and should be repaired immediately.

9. Test with a Known Working Appliance:

- Known Working Appliance: If you've checked all the wiring and connections and are still unsure, plug a known working appliance into the socket-outlet. If it operates correctly, the fault is likely with the original appliance rather than the circuit. This will rule out the possibility of the problem being with the device itself.

***Summary*:**

Diagnosing a "Loss of Power" situation involves checking for simple issues like an operated protective device or a switch left in the off position, and then progressively moving through the system with more detailed tests, such as voltage verification, continuity tests and inspecting for broken or loose connections. Always ensure that you are adhering to safety procedures while diagnosing power issues, including using appropriate PPE and testing equipment.

Protective Device Operating

When a protective device operates, it's a sign that the protective device is operating to protect the electrical installation and users from a potentially dangerous condition. These devices are designed to trip for several key reasons. Here's a guide to identify and troubleshoot the most common causes.

1. Overload

- What happens:
- The total load connected to the circuit exceeds the rating of the protective device.
- Signs:
  - Protective device trips after using multiple devices (e.g., toaster, kettle, microwave).
  - Also check the fridge socket-outlet, this can be one that is often forgotten as it is not easy to see.
  - Breaker feels warm to the touch.
- How to diagnose:
  - Unplug some devices.
  - Restore the protective device and test by turning appliances on one by one.
- How to fix:
  - Reduce load (move devices to different circuits).
  - If consistent, recommend installing a dedicated circuit.

2. Short Circuit

- What happens:
- A line conductor comes into contact with a neutral conductor or an earthed conductive part, creating a very low resistance fault path and a high fault current.

- Signs:
    - protective device trips immediately when the appliance is turned on.
    - Burning smell near socket-outlets or appliances.
    - Scorch marks on plugs or socket-outlets.
- How to diagnose:
    - Visually inspect socket-outlets, switches, appliances and accessible wiring.
    - Use insulation resistance testing if needed.
- How to fix:
    - Identify and isolate the fault.
    - Repair damaged wiring or replace faulty devices.
    - Repair or replace the damaged cable, accessory, or equipment before restoring the supply.

3. Earth Fault (RCD/RCBO operation)

- What happens:
- Current leaks from a line conductor to earth, often through a faulty appliance, moisture ingress or wiring.
- Signs:
    - RCD/RCBO trips randomly, sometimes without load increase.
    - Tripping when touching appliances or during wet conditions.
- How to diagnose:
    - Use an insulation resistance tester.
    - Disconnect suspect appliances and retest.
- How to fix:
    - Repair or replace faulty appliances or wiring.
    - Repair the fault and verify the operation of the RCD or RCBO before returning the circuit to service. Where residual current protection is not provided but is required by BS 7671, recommend upgrading the installation.

4. Nuisance Tripping

- What happens:
  - The RCD, RCBO, or MCB operates intermittently without an obvious fault being present. This may be caused by cumulative earth leakage, loose connections, moisture ingress, damaged insulation, or a developing fault that only occurs under certain operating conditions.
- Signs:
  - The protective device operates without an obvious overload or short circuit.
  - Happens intermittently.
- How to diagnose:
  - Tighten all terminal connections.
  - Disconnect loads or final circuits, one at a time.
  - Carry out insulation resistance testing.
- How to fix:
  - Repair the underlying fault. Replace the protective device only if testing confirms it is defective.
  - Correct any loose connections.

Pro Tips:

- Always check the simplest causes first (e.g., overloaded circuits).
- Remember to test under controlled conditions, always isolate power where necessary.
- Don't assume the protective device is faulty. In most cases, nuisance tripping is caused by the installation or connected equipment rather than the device itself.

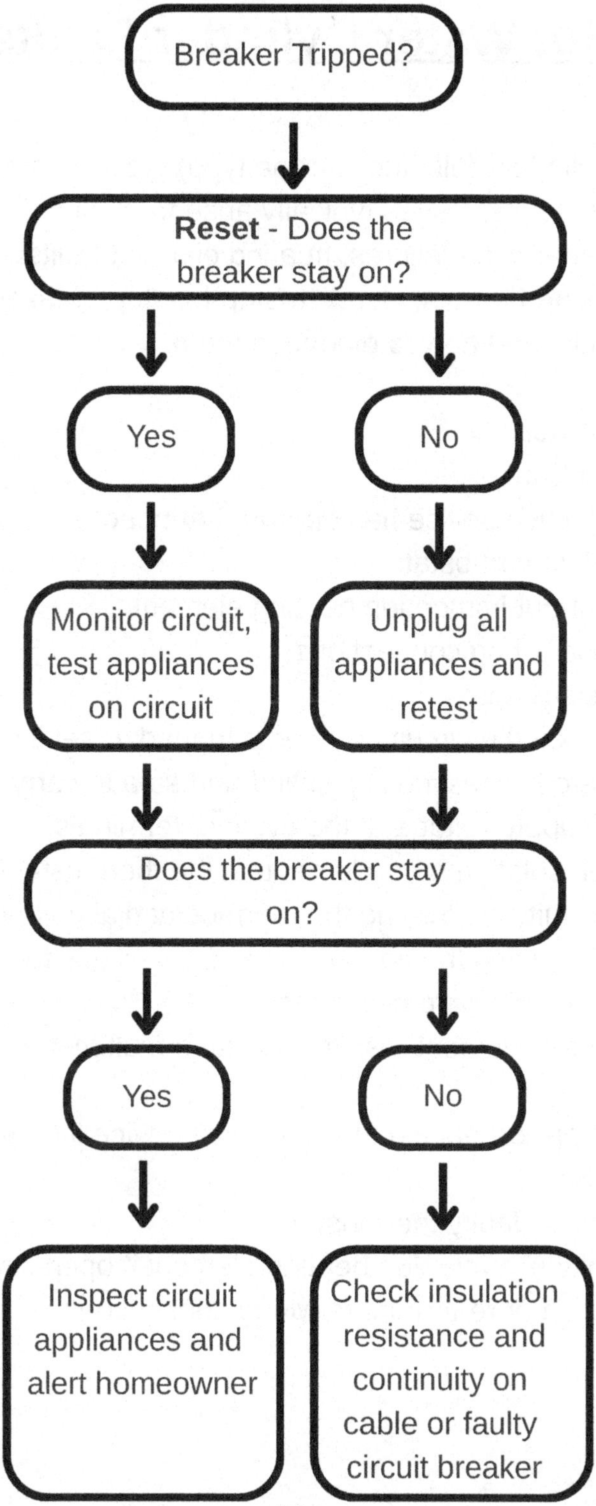
Breaker Tripped?
Reset - Does the breaker stay on?
Yes
No
Monitor circuit, test appliances on circuit
Unplug all appliances and retest
Does the breaker stay on?
Yes
No
Inspect circuit appliances and alert homeowner
Check insulation resistance and continuity on cable or faulty circuit breaker

Hot Water Cylinder Faults

Hot water cylinders (electric storage type) are a common source of household faults. Issues typically arise from electrical problems, thermostat failures, heating element faults, or plumbing-related issues. Here's a step-by-step breakdown of common faults and how to diagnose them.

1. No Hot Water

- Possible Causes:
  - Protective device has operated (tripped).
  - Faulty thermostat.
  - Burnt-out immersion heating element.
  - Loose wiring connections.
- Diagnosis Steps:
  - Check if the circuit breaker is tripped, reset if necessary.
  - Where live testing is justified and safe to carry out, verify the supply voltage at the cylinder terminals.
  - After isolation procedures, test the thermostat for continuity, try turning the thermostat dial up and down while testing the continuity to assess if it is functioning correctly (isolate circuit first).
  - Test the element resistance with a multimeter.
- Fix:
  - Restore or replace the protective device if found to be faulty.
  - Replace faulty thermostat.
  - Replace immersion heating element if open-circuit.
  - Tighten or re-terminate wiring connections.

2. Water Too Hot

- Possible Causes:
  - Thermostat contacts stuck closed.
  - Incorrect thermostat setting.
  - Thermostat sensor detached from cylinder.
- Diagnosis Steps:
  - Inspect thermostat set temperature (typically should be set around 60°C).
  - Test the thermostat to confirm it opens and closes at the correct temperature.
- Fix:
  - Adjust thermostat setting.
  - Replace thermostat if stuck closed.

3. MCB,RCBO or RCD Operating

- Possible Causes:
  - Shorted heating element (internal short to earth).
  - Moisture ingress into cylinder wiring or terminal block.
  - Damaged wiring insulation.
  - Failed immersion heater element leaking to earth when hot.
- Diagnosis Steps:
  - Carry out an insulation resistance test between line and earth, and neutral and earth.
  - Inspect element gland and wiring.
- Fix:
  - Replace faulty element.
  - Repair/replace damaged wiring.
  - Seal or replace leaking glands or cylinder fittings.

4. Water Leaking from the Cylinder

- Possible Causes:
  - Immersion heater element is loose, or the sealing gasket is damaged or missing.
  - Faulty pressure relief valve (PRV) discharging constantly.
  - Loose pipework fittings.
- Diagnosis Steps:
  - Identify the source of the leak. If the leak is from pipework, the cylinder, or a plumbing fitting rather than the immersion heater assembly, refer the repair to a competent plumber or heating engineer.
- Fix:
  - Isolate and drain water from cylinder, unscrew and check for seal on immersion heater, replace seal if needed and retighten.

5. Slow Heating / Insufficient Hot Water

- Possible Causes:
  - Partially failed heating element (low output).
  - Incorrect thermostat setting.
  - Limescale or sediment build-up in cylinder
- Diagnosis Steps:
  - Test element resistance (compare to rated value).
  - Check thermostat setting.
  - Consider age of cylinder and water hardness in the area.
- Fix:
  - Replace faulty element.
  - Adjust thermostat.
  - Flush cylinder if heavily scaled (plumber usually needed).

Pro Tip: Crack the immersion heater slightly loose before draining the cylinder. If you wait until the cylinder is empty, it can be incredibly difficult to remove, especially on older installations where the immersion heater has seized in place.

How to calculate an expected element resistance

To check if a hot water element is within range, you can use Ohm's Law. Start with the power rating on the nameplate.

Example (3.0 kW element at 230 V):
I = Current (Amps)
P = Power (Watts)
V = Voltage (Volts)
R = Resistance (Ohms)

1. Calculate the Amps from the Watts.
$I = P/V = 3000/230 = 13.04A$

2. Now calculate the Resistance.
$R = V/I = 230/13.04 = 17.7\Omega$

So a 3.0 kW element should measure around 17–18 Ω cold. Allow ±10% for tolerance.

Open circuit
If the element has no continuity (open circuit or OL on the meter), the heating element has failed and should be replaced.

Resistance within range
If the measured resistance is close to the calculated value (typically within ±10%), the heating element is likely electrically sound. An insulation resistance test should also be carried out to confirm there is no leakage to earth.

Resistance outside expected range
If the measured resistance differs significantly from the calculated value, confirm the element's power rating first. If the rating is correct, the element may be damaged.

4

COMMERCIAL FAULTS

Power Distribution Systems in Commercial Buildings

Overview

Commercial buildings often have complex power distribution systems involving multiple distribution boards, and both single-phase and three-phase circuits. These systems are designed to handle high and variable loads such as HVAC systems, lifts, lighting, and office equipment. A fault in the distribution system can affect large areas and critical services, so efficient fault-finding is essential.

Common Faults in Power Distribution Systems

1. Protective Devices Operating

- Symptoms: MCB,RCD/RCBO repeatedly trips under load.
- Potential Causes:
    - Overloaded circuit due to added equipment.
    - Nuisance tripping from cumulative earth leakage.
    - Faulty appliances or damaged cabling downstream.
    - Neutral-to-earth faults or short circuits.
    - Loose terminations caused by thermal cycling.
- Rectification Steps:
    - Identify what's connected to the circuit. Unplug devices one by one.
    - Use a clamp meter to check current draw, compare to protective device rating.
    - Perform insulation resistance testing to locate damaged cable or appliance.
    - Check neutral-earth paths for bonding or wiring issues.

2. Phase Imbalance

- Symptoms:
  - Equipment running inefficiently, protective device unusually warm, excessive humming.
- Potential Causes:
  - Uneven loading across phases, especially in three-phase boards.
  - Single-phase loads not balanced.
  - Faulty equipment drawing excessive current from one phase.
- Rectification Steps:
  - Measure the current on each phase using a clamp meter during periods of peak demand to identify any imbalance.
  - Re-balance loads where possible by relocating circuits.
  - Review the distribution board schedule and redistribute single-phase circuits where practical to balance the load.

3. Voltage Drop or Fluctuation

- Symptoms: Lights dimming, motors stalling, equipment resets.
- Potential Causes:
  - Undersized cables for load distance.
  - Loose terminations or corroded connections.
  - Faulty main switchgear or incoming supply issues.
- Rectification Steps:
  - Measure the voltage at the origin of the circuit and at the point of utilisation while the circuit is under load. Compare the readings to determine the voltage drop.
  - Tighten or replace loose/burned connections.
  - Where appropriate, verify the incoming supply and earthing arrangement. If a supply fault is suspected, report it to the Distribution Network Operator (DNO).
  - Where voltage drop exceeds the limits permitted by BS 7671, consider increasing the conductor size.

4. Neutral Faults

- Symptoms: Appliances damaged, lights over-bright or too dim, erratic behaviour.
- Potential Causes:
  - Open or floating neutral on the distribution system.
  - Loose or broken neutral link at the board.
  - Neutral conductor damaged or corroded.
- Rectification Steps:
  - Test for voltage between line and neutral at multiple points.
  - Check continuity of neutral from source to load.
  - Inspect distribution boards for loose terminations or signs of heat.
  - Always isolate before tightening or replacing neutral conductors.

5. Distribution Board Faults

- Symptoms: No power to circuits, protective devices failing to reset, buzzing sounds.
- Potential Causes:
  - Loose connections or thermal damage at main busbars or protective devices.
  - Mechanical failure of circuit breakers.
  - Arcing or water ingress into the board.
- Rectification Steps:
  - Visually inspect for burn marks, corrosion, and signs of water entry.
  - Inspect suspect protective devices for mechanical or thermal failure. Replace if necessary.
  - Re-torque accessible terminations to the manufacturer's specified tightening torque where required.
  - If corrosion or water is found, the distribution board may need replacement, repair or rebuild.

Tools Recommended

- Insulation resistance tester
- Clamp meter
- Multimeter (true RMS preferred)
- Thermal imaging camera (for hotspots)
- Torque screwdriver (for tightening terminations)
- Voltage drop calculator (or app)
- Multifunction Installation Tester (MFT)

Pro Tip:

In commercial installations, many faults are the result of previous alterations rather than the original installation. Always look for undocumented modifications, undersized conductors, overloaded circuits, or poorly terminated cables. Before diving into detailed testing, ask: "Has anything been changed recently?"

HVAC Systems: Diagnosing Faults

Overview

HVAC (Heating, Ventilation and Air Conditioning) systems in commercial environments integrate low-voltage controls (often 24V AC), sensors, relays, contactors, and high-voltage three-phase motors for compressors and fans. Faults can occur anywhere along this chain, from a faulty thermostat sensor to a failed three-phase condenser fan. Understanding the control logic and signal flow is critical for fast, effective troubleshooting.

Common HVAC Fault Areas & Fault-Finding Steps

1. Thermostats & Temperature Sensors

Symptoms:

- Room not reaching set temperature.
- HVAC not turning on or off.
- Inaccurate temperature readings.

Potential Causes:

- Faulty thermistor, temperature sensor or RTD.
- Broken wires or poor terminations.
- Incorrectly placed sensor (in draft or sun).

Rectification:

1. Measure resistance of the probe (e.g., 10kΩ at 25°C for common thermistors).
2. Verify the controller is supplying the correct control voltage (commonly 24 V AC) to the thermostat or sensor circuit.
3. Compare to manufacturer's resistance chart.
4. Check wiring continuity back to the controller.
5. Relocate sensor if environmental placement is skewing readings.

2. 24V AC Control Circuit (Low-Voltage Controls)

Symptoms:

- No response from unit.
- Relays or contactors not pulling in.

Potential Causes:

- Blown control fuse.
- Open control transformer secondary.
- Broken or damaged control cable (commonly on roof runs).
- Faulty relay or shorted coil.

Rectification:

- Check for 24V AC between R and C terminals at thermostat or control board.
- Test control fuse continuity.
- Inspect transformer, confirm primary and secondary voltages.
- Check resistance across relay coils, replace if open.
- Look for shorts between 24V wires and earth.

3. Contactors & Relays

Symptoms:

- Indoor or outdoor fans not starting.
- Compressor doesn't run.
- Loud buzzing from contactor.

Potential Causes:

- Coil not energising.
- Burnt contacts.
- Mechanical failure (sticking or chattering).

Rectification:

- Confirm correct control voltage is present (24V or 230V, depending on design).
- Where it is safe to do so, manually operate the contactor to confirm the load operates correctly.
- Inspect contacts for pitting or carbon buildup.
- Replace faulty contactor or relay.

4. Condenser & Evaporator Fan Motors

Symptoms:

- Overheating or thermal overload operating
- No airflow.
- Fan hums but doesn't spin.

Potential Causes:

- Seized bearings.
- Failed start/run capacitor.
- Open windings.
- Phase loss on 3-phase fans.
- Blocked air filter.

Rectification:

- Isolate supply, spin fan blade manually, should rotate freely.
- Test capacitor with capacitance meter.
- Test windings for continuity and insulation resistance.
- On 3-phase fans, test voltage between each phase, check for imbalance or phase loss.
- Replace motor or capacitor as required.
- Clean or replace blocked filters, remove any debris.

5. Compressors (Scroll or Reciprocating, 3-Phase)

Symptoms:

- No cooling.
- Tripping breaker.
- Compressor starts then shuts off.

Potential Causes:

- Open windings or internal overload.
- High head pressure (blocked condenser).
- Start/run capacitor fault (if single phase).
- Phase reversal.
- Locked rotor.
- High-pressure or low-pressure safety switch operating.

Rectification:

- Isolate the supply, measure resistance between terminals (C-R-S) and check for open circuit or shorts to earth.
- Test phase rotation using a phase sequence meter.
- Ensure correct voltage is supplied to all three phases.
- Check pressure switches and high-pressure cut-out for resets.
- If windings test OK but compressor won't start, possible mechanical failure (replace unit).

6. Pressure Switches and Safety Devices

Symptoms:

- Unit locked out.
- No compressor activity despite good control signal.

Potential Causes:

- Low or high refrigerant pressure.
- Faulty switch.
- Tripped manual reset.
- Refrigerant gas leak.

Rectification:

- Check refrigerant pressures (requires appropriate license).
- Confirm switch continuity, test between COM and NC/NO.
- Reset if it's a manual switch and safe to do so.
- Investigate cause of trip, blocked airflow, low charge, etc.

7. Control Boards & Fault Codes

Symptoms:

- LED flashing codes.
- Random shut-downs
- No response to commands.

Potential Causes:

- Control board failure.
- Sensor feedback error.
- Software lockout due to repeated trips.

Rectification:

- Read the fault code and refer to the manufacturer's diagnostic chart.
- Power-cycle the system and monitor for reoccurrence.
- Inspect the control board for damaged components, dry solder joints, burnt tracks, swollen capacitors, or signs of overheating.
- Replace the control board if diagnostics confirm failure.

Pro Tip:

If the compressor isn't running, don't assume it's faulty. Start at the control circuit and work your way towards the high-voltage components. Verify the control voltage, safety switches, contactor operation, supply voltage, and phase rotation before condemning the compressor.

Real-World Fault Example -Rooftop HVAC Not Working

Scenario:

After a heavy storm, a rooftop HVAC unit stopped working. The indoor thermostat appeared normal and responsive, but the outdoor unit was completely inactive, no fan, no compressor, no sound.

Initial Checks:

Power to the building was confirmed, and the main isolator feeding the HVAC system was on. No visible signs of damage at the outdoor unit.

Testing:

Checked for 24V AC at the control board, nothing present. This indicated the control circuit wasn't being powered.

Diagnosis:
Upon inspection, the low-voltage control fuse on the board was found blown. Further testing revealed a short in the control wiring running to the outdoor coil, likely due to storm-related water ingress or physical damage. Additionally, the relay contactor coil showed signs of thermal damage, possibly caused by the short circuit or transient voltages during the storm.
Fix:

- Replaced the damaged control wiring.
- Installed a new control fuse.
- Replaced the burnt-out contactor.

Outcome:
Once power was restored, the system resumed normal operation. Cooling returned, and the thermostat responded correctly. The system completed several operating cycles without any further faults.

Key Lesson:
Always verify low-voltage control power when a unit appears dead. A blown control fuse often points to a shorted external device or wire, and storms frequently cause this type of fault due to water, debris, or voltage surges.

Emergency Lighting Faults

Overview

Emergency lighting systems are critical for occupant safety during loss of the normal supply, providing illumination for safe evacuation. These systems must remain functional even when normal lighting circuits fail. In commercial environments, faults in emergency lighting often go unnoticed until routine inspection and testing, or a real emergency, exposes them.

Emergency lighting typically includes:

- Batteries or central battery systems.
- Self-contained battery-backed emergency luminaires.
- Test switches.
- Exit signs and emergency luminaires.
- Integration with normal lighting circuits for charging and monitoring.

Common Emergency Lighting Faults

Battery Failures

- Symptoms: Emergency light does not stay on during outage or self-test.
- Causes: Old or degraded batteries, failed charging circuit, or simply the battery has not been connected.
- Fix: Replace battery packs, test charging output, ensure battery is connected.

Lamp or LED Failure

- Symptoms: The unit powers on but provides no illumination.
- Causes: LED driver failure, blown lamp, heat damage.
- Fix: Replace failed components; test voltage at lamp terminals.

Charger Circuit Faults

- Symptoms: Battery not holding charge or overcharging.
- Causes: Failed capacitors, burnt PCB tracks, charger circuit fault.
- Fix: Repair PCB or replace fitting if not serviceable.

Test Button or Key Switch Not Working

- Symptoms: Light won't activate in test mode.
- Causes: Faulty switch, broken test circuit wiring, controller fault.
- Fix: Test continuity across switch, trace wiring path, verify correct voltage is present.

Loose or Missing Neutral

- Symptoms: Some lights fail to charge or don't function during test.
- Causes: Poor connections at terminal blocks, shared neutrals disconnected.
- Fix: Inspect terminations in distribution boards and junction boxes; test continuity between line and neutral.

Exit Signs Not Illuminating

- Symptoms: Sign is dark during test or normal operation.
- Causes: Power supply issue, PCB or LED failure, incorrect wiring during renovations.
- Fix: Confirm voltage supply; test internal battery; inspect driver circuitry.

Real-World Example: Emergency Luminaires Failing Routine Test

Fault Description:
During a scheduled 3-hour emergency lighting discharge test in a commercial office block, multiple emergency luminaires and exit signs across one floor failed to illuminate. Some lit dimly, while others stayed off entirely. A facilities manager requested an urgent inspection before an upcoming compliance audit.

Step-by-Step Walkthrough Using the Fault-Finding Process:
1. Initial Assessment
I spoke with the building manager and reviewed the site log. They mentioned:

"We had LED downlights replaced recently in the same zone."
"The fault only affects one area, the rest of the building passed the test."

This pointed to a localised fault, possibly introduced during recent lighting alterations.

2. Safety Checks

- Completed the safe isolation procedure before working on any exposed conductors.
- Checked for backfeed or shared supplies.

3. Diagram Analysis

- No updated as-built drawings were available, but label IDs on the distribution board indicated multiple circuits tied into an emergency test key switch.
- Traced affected circuits physically using core numbers and verified test key input back to emergency luminaires.

4. Testing

- Voltage test on live lighting circuits showed 230V present.
- An abnormally high voltage was measured between neutral and earth, indicating a possible open or high-resistance neutral.
- Continuity test on neutral confirmed an open circuit between the distribution board and emergency luminaires.

5. Diagnosis

- A neutral conductor in the distribution board was loose, only partially clamped. Likely dislodged when new lighting was installed.
- This caused incomplete charging on the emergency light circuit and unstable performance during discharge.

6. Repair

- Fully isolated the board and re-terminated the neutral using proper torque settings. Checked all neutral terminations.
- Retested continuity and voltage, confirmed solid connections.

7. Final Check

- Performed a functional emergency luminaire test, all fittings now illuminated and discharged correctly.
- Informed client of the cause, the fix, and recommended a re-test before audit day.
- Logged repair in maintenance schedule and added a note about as-built documentation being outdated.

Key Lesson:

Never assume lights are faulty until the circuit's integrity is verified. A loose neutral, especially on shared control circuits, can cause scattered faults that look like equipment failure but are really just poor terminations.

5

INDUSTRIAL FAULTS

Motor Control Circuits and Faults

Overview

Motor control circuits in industrial settings can range from basic direct-on-line (DOL) starters to advanced VSD (Variable Speed Drive) and PLC-controlled systems. These circuits handle high loads and operate in harsh conditions, so faults are not uncommon. In most motor control faults, start by verifying the incoming supply, then diagnose the control circuit before moving on to the power circuit. Many "motor won't start" faults are caused by issues within the control circuit rather than the motor.

Common Motor Control Circuit Faults

1. Motor Won't Start

Symptoms:

- No movement or sound from the motor.
- Control buttons (Start/Stop) seem unresponsive.
- No contactor activity.

Potential Causes:

- Blown control fuse.
- Open stop button or safety interlock.
- Failed start button.
- Coil voltage not reaching contactor.

Rectification Steps:

1. Verify incoming supply is present.
2. Test start/stop circuit for continuity and correct operation.
3. Measure voltage at contactor coil, should match control voltage rating.
4. Check overload relay for trip condition, identify the cause and reset if needed.
5. Inspect safety circuits (e.g., E-stops, guards) for breaks.

2. Contactor Buzzing or making unusual noise

Symptoms:

- Loud buzzing from control panel.
- Contactor won't latch.
- Motor stutters or cycles on/off.

Potential Causes:

- Low coil voltage (weak supply or long run).
- AC coil on DC circuit (or vice versa).
- Mechanical binding in the contactor (stuck).
- Control voltage drop during energisation.

Rectification Steps:

- Measure coil voltage under load, should be within ±10% of rated value.
- Confirm correct coil type is installed (AC vs. DC).
- Replace worn or stuck contactors.
- Check wire size and connections for voltage drop issues.
- Isolate supply and use contact cleaner to remove friction, ensure contact is dry before energising.

3. Motor Trips Immediately on Start

Symptoms:

- Protective device or overload trips.
- Motor hums but doesn't turn.
- Excessive inrush current.

Potential Causes:

- Locked rotor or seized bearings.
- Incorrect phase rotation.
- Shorted windings.
- Undersized protection settings.

Pro Tip:

Always check off the basics first, something that might seem overlooked could be the fault. Motor faults often begin with something small.

Rectification Steps:

- After isolating the circuit, try rotating shaft manually, should move freely.
- Perform insulation resistance test (Megger) to check for shorts.
- Verify the protective device and overload relay are correctly selected and adjusted for the motor's full-load current.
- Use clamp meter during startup to observe inrush.
- Check supply for all three phases, phase loss or imbalance?

4. Motor Runs But Overload Trips After a While

Symptoms:

- Motor starts normally, then trips after a few minutes.
- Overload reset required.
- Motor hot to touch.

Potential Causes:

- Overloaded mechanical load.
- Fan cooling system blocked or failed.
- Incorrect overload setting or curve.
- Voltage imbalance or single-phasing.

Rectification Steps:

- Compare actual current draw to motor nameplate and overload setting.
- Inspect load mechanically (e.g., jammed pump or conveyor).
- Test phase-to-phase voltage for imbalance.
- Clean motor vents and check external fan function.

5. Start/Stop Buttons Not Working Properly

Symptoms:

- Motor won’t respond to button presses.
- Only works when holding button down.
- Intermittent operation.

Potential Causes:

- Faulty or sticky pushbutton.
- Loose wiring.
- Control circuit not latching (holding contact open).

Rectification Steps:

- Test button operation with continuity tester.
- Inspect terminal blocks for loose or corroded wires.
- Check holding contact in the contactor circuit, must bridge when energised.
- Confirm control transformer and fusing are intact.

6. Overload Relay False Tripping

Symptoms:

- Tripping with no apparent cause.
- Happens under light load.
- Trips inconsistently.

Potential Causes:

- Ambient temperature affecting thermal type overload.
- Loose motor terminals causing high resistance heating.
- Incorrect settings on overload.

Rectification Steps:

- Replace thermal with electronic overload for better accuracy.
- Tighten all terminal screws with torque driver.
- Compare overload type to application (e.g., conveyor vs. fan).
- Consider external ambient compensation if control panel gets hot.

Tools Recommended:

- Multimeter (for voltage and continuity checks).
- Clamp meter (to check motor current draw).
- Insulation resistance tester (for winding checks).
- Torque screwdriver (for terminal checks).
- Control circuit schematic (if available).
- Phase sequence meter.

Control Panels and PLC Systems

What is a Control Panel?

An industrial control panel is an enclosure housing electrical and electronic components used to control machinery, processes, and safety systems in a plant. It typically includes:

- Power distribution equipment (protective devices, isolators)
- Motor starters (DOL, star-delta, VSDs)
- Control relays, timers, and contactors
- Programmable Logic Controllers (PLCs)
- Safety devices (E-stops, safety relays, interlocks)
- Terminal strips and labelled wiring for field connections

These control panels are built for reliability and safety, often with internal segregation for different voltages (e.g., 24VDC control vs 400VAC power).

What is a PLC?

A Programmable Logic Controller (PLC) is an industrial computer used to automate and control machines or processes. It monitors inputs, executes programmed logic, and switches outputs accordingly.

Think of it as the brain of the system, reading sensors and sending commands to actuators.

PLC Inputs and Outputs (I/O)

<u>Inputs:</u>

Inputs tell the PLC what's happening in the real world. These can be:

- Digital Inputs (DI): ON/OFF signals
  - Examples:
    - E-stop pressed
    - Limit switch triggered
    - Start/Stop button pressed

- Analog Inputs (AI): Variable signals
  - Examples:
    - Temperature transmitter (0–10V or 4–20mA)
    - Pressure sensor
    - Flow meter

The input terminals are **usually** 24VDC and go into I/O modules on the PLC.

Outputs:

Outputs tell field devices what to do. These can be:

- Digital Outputs (DO): Activate or deactivate devices
  - Examples:
    - Start a motor via contactor
    - Turn on a pilot light
    - Open/close a solenoid valve
- Analog Outputs (AO): Send variable control signals
  - Examples:
    - Adjust VSD speed
    - Control a valve position

Safety Relays and Interlocks

A safety relay is a fail-safe device that ensures safety-critical functions (like E-stops, guards, and light curtains) are properly executed.

Key points:

- Monitors circuits like emergency stops and safety doors.
- Provides dual-channel redundancy to prevent single-point failures.
- Must be reset manually after the fault is cleared.
- Interrupts power to hazardous outputs (like motors or hydraulics).

You'll typically find:

- Dedicated safety circuit terminal blocks
- LED indicators showing status of safety functions

Component	Description
Main Isolator	Shuts off power to control panel for safe access
Control Transformer	Steps down 400VAC to 24VDC/AC for control circuit
PLC Rack	Modular I/O (input/output) cards, CPU and often labelled correctly
Wiring Ducts/Trunking	Contains cables neatly, may contain cable labels within here
Terminal Strips	Connects field devices to control components
Safety Relays	Brightly coloured (often red/yellow) and controls E-stop devices
Wiring and Device Identification	Each wire and device should have a number relating to the schematic
HMI (Human Machine Interface)	This screen is to view and control the processes

Pro Tip:

Always look for the LED status indicators on I/O cards. These give immediate clues: input on, output active, or fault present.

PLC Faults

When a machine or process controlled by a PLC isn't working correctly, the fault may not be with the mechanical components, but with the PLC, I/O modules, wiring, or logic. Below are common PLC-related faults and how to systematically troubleshoot them.

1. No Power to PLC (PLC Not Starting)

Symptoms:

- No lights on PLC CPU.
- No response from HMI.
- Machine appears completely dead.

Diagnosis:

- Verify control voltage at PLC power terminals (typically 24VDC).
- Inspect upstream fuses, breakers, or power supply module.
- Use multimeter to confirm power supply is present and within the specified operating range.

Fix:

- Restore blown fuse or reset breaker after rectifying the cause.
- Replace faulty PSU or PLC power module.
- Ensure all terminations are tight and corrosion-free.

2. PLC in Fault Mode (Flashing/Error Light)

Symptoms:

- CPU has a red or amber LED.
- HMI may show fault message or lose communication.

Diagnosis:

- Contact available personnel who is proficient in the software who can access the PLC.
- Read error or diagnostic code.
- Common causes: corrupted program, memory full, hardware fault, watchdog timeout.

Fix:

- Re-upload valid program from backup.
- Clear fault and reload.

3. Input Not Registering

Symptoms:

- Sensor/switch is live, but PLC doesn't respond.
- Input LED on PLC module remains off.

Diagnosis:

- Confirm field device is working and receiving correct voltage.
- Measure voltage at PLC input terminal.
- Check wiring for damage, loose terminals, or broken conductors.

Fix:

- Repair wiring or replace failed sensor.
- If input is present but LED still off, replace input module.
- Check common terminals, PLC input modules often require a shared return.

4. Output Not Working

Symptoms:

- Output LED on PLC is on, but the device doesn't activate (e.g., motor won't start).
- Solenoids or contactors don't respond.

Diagnosis:

- Measure voltage at PLC output terminal when LED is on.
- Check if fuse on output module is blown.
- Inspect field wiring and device (e.g., motor contactor coil).
- Inspect the relay contacts for excessive wear or burning and verify continuity when the relay is energised.

Pro Tip: Always follow the safe isolation procedure and prove dead before working on electrical equipment. Never rely on indicator lights or the HMI to confirm a circuit is de-energised.

Fix:

- Replace blown fuse or faulty output module after rectifying the fault.
- Replace burnt relay or solid-state output.
- Repair damaged field wiring or replace actuator.

5. Program Logic Faults

Symptoms:

- Inputs and outputs are fine, but machine behaves incorrectly.
- Timers or sequences out of sync.
- HMI controls not responding as expected.

Diagnosis:

- Contact suitable personnel who is proficient in the software who can access the PLC.
- Read error or diagnostic code.

Fix:

- Correct logical errors in the program.
- Re-download updated or backup version of correct logic.
- Check timers, counters, memory registers, and internal bits for unexpected values or incorrect operation.

6. Communication Loss (PLC ↔ HMI or Network Devices)

Symptoms:

- HMI is blank or says “comm error.”
- Remote I/O or drives stop responding.
- SCADA can’t access PLC.

Diagnosis:

- Check network cables and connections.
- Look for switch or router faults.
- Use ping or software tools to test connectivity.
- Check IP address conflicts or mismatches.
- Inspect CPU or communication module status LEDs.

Fix:

- Reset or reconfigure network hardware.

- Replace failed communication module.
- Reconfigure IP addresses if duplicated.

7. Faulty Safety Inputs or Relays

Symptoms:

- E-stop won't reset.
- Safety light curtain blocks machine from starting.
- Safety relay shows fault indicator.

Diagnosis:

- Inspect status LEDs on safety relay (often yellow/red).
- Verify both safety channels are operating correctly and in agreement, as required by the safety relay design.
- Confirm all guards/safety devices are closed or unblocked.
- Look for earth faults or feedback wiring errors.

Fix:

- Repair faulty E-stop or door switch.
- Only reset the safety relay after confirming all safety conditions have been restored.
- Replace safety relay if internal fault detected.

General Tips for PLC Troubleshooting

- Label Everything: Proper labelling makes tracing faults much easier.
- Use Schematics: Compare control panel wiring to diagrams to trace I/O connections.
- Record Changes: Log all program updates or field wiring modifications.
- Look at LEDs: They're your first line of defence in fault diagnostics.

<u>Pro Tip:</u>

Start with the PLC status LEDs. Confirm the PLC has power, the safety circuit is healthy, and the expected inputs are being received before investigating the outputs.

Testing Faulty Sensors

Industrial sensors such as pressure, temperature, flow, and level transmitters often output a 4–20 mA signal, which represents a linear reading. These analog loops are widely used because they're resistant to electrical noise and are highly resistant to electrical interference and can be transmitted over long cable runs.

Common Sensor Fault Symptoms:

- Display shows 0.0, "out of range," or "sensor error."
- HMI or PLC tags are frozen or fluctuating.
- The process reacts incorrectly or goes into alarm mode.
- Field devices don't respond to changes in process conditions.

How to Test 4–20 mA Sensors

Step 1: Visually Inspect Wiring

- Check terminations are tight and correct.
- Look for broken, burnt, or corroded wires.
- Ensure correct polarity (especially for loop-powered devices).

Step 2: Measure the Loop Current

Tool Needed: 4-20mA process clamp meter.

Method 1: Using a clamp meter on mA range

- Use a loop clamp meter (e.g., Fluke 773) around a single conductor of the loop.
- Read the current directly, normal range is 4.00 to 20.00 mA.

Method 2: Using a multimeter in series

- Break the loop circuit.
- Place multimeter leads in series with the sensor and power supply.
- Set the meter to mA DC and read the current.

Testing a Sensor

Current	Condition
<4mA	Sensor fault, open loop or wiring issue
4mA	Sensor powered but measuring zero
12mA	Normal mid scale reading
20mA	Full scale reading
>20mA	Possible short, over-range or internal fault

Step 3: Simulate or Substitute Signal

- Use a mA loop calibrator to simulate 4–20 mA signal and confirm the receiving device (e.g., analogue input module) is working.
- If you have a known good spare sensor, temporarily install it to test the loop.

Step 4: Check Sensor Supply Voltage

- Some sensors require 24VDC or loop power.
- Test across supply terminals with a multimeter.
- Low or missing voltage could mean power supply failure or wiring issue.

Testing Resistance-Type Probes (RTDs, Thermistors)

For non-loop sensors like RTDs and thermistors:
Disconnect and Test Resistance

- Use a multimeter on ohms setting.
- Measure across sensor terminals.

<u>Sensor Type</u>	**<u>Expected Range</u>**	**<u>Open Circuit</u>**	**<u>Short Circuit</u>**	**<u>Notes</u>**
PT100RTD	100 Ω at 0°C	Infinite Ω (Broken sensor or damaged wire)	Near 0 Ω (Broken sensor or damaged wire)	Resistance increases with temperature
NTC Thermistor	10k Ω at 25°C	Infinite Ω (Broken sensor or damaged wire)	Near 0 Ω (Broken sensor or damaged wire)	Resistance decreases with temperature

<u>Pro Tip:</u>

Always test from the PLC or controller end first.

If you're reading 0 mA or a fault at the input terminal, disconnect the sensor and simulate a 4–20 mA signal with a calibrator or test loop.

If the PLC responds correctly, your sensor or field wiring is the issue. This isolates the fault quickly without pulling out every device in the loop.

Troubleshooting Heavy Machinery

Overview

Heavy machinery in industrial settings includes equipment like conveyors, crushers, pumps, compactors, and mobile plant. These machines are typically powered by large motors, hydraulics, and control systems, often integrating PLCs, VSDs (variable speed drives), safety relays, sensors, and HMIs. Heavy machinery faults can cause major downtime and safety risks, so accurate, efficient fault-finding is crucial.

Step-by-Step Fault-Finding Process

1. Visual and Safety Inspection

- Safely isolate the equipment and apply lock-off devices before carrying out any work. Follow the site's safe isolation procedure.
- Check for: loose wires, blown fuses, oil leaks, burnt components, stuck contactors, damaged cables or emergency stop engaged.
- Make sure all guards are in place and no safety interlocks are open.

2. Check Power Supply

- Test incoming three-phase voltage at main terminals (L1-L2-L3) using a multimeter.
- Look for:
    - Imbalance between phases
    - Voltage drop under load
    - Loose terminals (especially in vibration-prone environments)

3. Test Control Circuit

- Use a control circuit diagram if available.

- Confirm:
  - 24VDC or 110/240VAC control power is present.
  - Emergency stops and safety relays are latched.
  - Start/stop pushbuttons are working and returning correct voltage.
  - PLC or relay logic is sequencing properly (use indicators or PLC status LEDs).

4. Check Motors and Drives

- Motor won't run?
  - Test coil of contactor (does it energise?).
  - Measure resistance of motor windings (test phase-to-phase and phase-to-earth).
- If VSD is installed:
  - Look at the fault code on display.
  - Test for correct enable signal and reference input.
  - Confirm output voltage and frequency to motor.

5. Sensor and Field Device Testing

- Test all limit switches, pressure sensors, encoders, and proximity sensors.
- Use a multimeter to check voltage change on actuation.
- For analog devices, test for 4–20 mA or 0–10 V output.

6. Use Bypass/Test Mode (if safe to do so)

- If the machine has a "test" or "jog" function, use it to isolate sections.
- Temporarily bypass sensors to confirm logic response, only if it doesn't compromise safety and is authorised on site.

7. Check Hydraulic or Pneumatic Systems

- Look for stuck valves, low pressure, or air leaks.
- Use pressure gauges to test system health.
- Check if solenoids are energising when commanded.

Example Fault Scenario: Conveyor Belt Will Not Start

System Setup:

- 3-phase conveyor motor controlled via VSD
- Start/Stop buttons at local station
- Safety circuit with E-Stop and belt misalignment limit switch
- PLC handling start signal logic and displaying status on HMI

Scenario:

A conveyor belt system driven by a 3-phase motor and controlled via a VSD (Variable Speed Drive) would not start. The operator attempted to start it both from the HMI and the local Start button, but nothing happened. The motor made no sound, and the contactor didn't pull in. The VSD displayed "Ready," but the motor wouldn't run.

Initial Checks:

- Verified 3-phase power present at the main isolator.
- VSD confirmed power was good by showing a "Ready" status.
- 24VDC control voltage was present at the PLC and HMI.

Testing:

Pressed the Start button, no 24V signal reached the PLC input that triggers the run command. Start pushbutton tested OK and was closing momentarily as expected.

Diagnosis:

Traced the Start signal path through the safety circuit. Found the belt misalignment (drift) limit switch was stuck in the open position, breaking the circuit. Emergency stop was checked and confirmed reset and latched.

Temporary Test:
As part of a controlled fault-finding procedure, and in accordance with the site's safe system of work, the belt drift switch circuit was temporarily simulated to verify the control logic. Pressing the Start button now sent the correct signal to the PLC. The VSD received the run command and the motor started as expected. The safety circuit was immediately returned to its original condition after testing.

Fix:

- Replaced the faulty belt drift limit switch.
- Adjusted switch to trigger properly when belt is aligned.
- Verified full operation of the safety circuit.

Outcome:
The conveyor system resumed normal operation. The start signal passed correctly through the safety chain, and the motor ran when commanded.

Key Lesson:
When a start command doesn't reach the PLC or VSD, always check the full path, especially safety devices like limit switches and E-Stops, which often interrupt control circuits silently.

Root Cause:
Belt had walked out of alignment and tripped the drift switch. The switch was not resetting due to mechanical wear, needed replacement.

Pro Tip:
Safety limit switches in heavy machinery are often the first thing to fail due to physical wear or impact. Always check mechanical actuation and electrical continuity when a machine won't start.

How to Test a Motor

Overview
Three-phase motors are the workhorses of the electrical industry, used in fans, pumps, conveyors, compressors, and processing equipment. Most faults fall into three main categories:

1. Electrical faults
2. Mechanical faults
3. Control/system faults

Knowing how to correctly test motors before removing them saves hours of labour and downtime.

Motor Testing Procedure

1. Visual Inspection

- Check nameplate for voltage, current, and wiring config
- Look for burnt smell, discolouration, cracked insulation, oil ingress

2. Supply Check

- Use multimeter or clamp meter to check:
    - Voltage on all 3 phases
    - Phase rotation if applicable
    - Current draw per phase

3. Insulation Resistance Test (Megger Test)

- Use 500V or 1000V insulation tester depending on the motor and installation.
- Test each winding to earth and winding-to-winding

Pro Tip: Never condemn a motor until you've verified the supply, control circuit, and driven load. Many "motor faults" are actually caused by problems elsewhere in the system.

Motor Insulation Resistance Testing Table

Test	***Result***
U Winding to Earth	>100 MΩ (ideal)
V Winding to Earth	>100 MΩ (ideal)
W Winding to Earth	>100 MΩ (ideal)

If readings are low (<1 MΩ), this will be due to water ingress or the winding insulation breaking down.

Pro tip:
Remove the motor terminal links before testing the windings. Leaving the links in place can produce misleading results by connecting multiple windings together during the test.

4. Winding Resistance Test

- Use a low-ohm meter (milli-ohm if possible) or multimeter.
- Compare U–V, V–W, and U–W.
- The winding resistances should be similar. A significant difference between phases may indicate a winding fault.
- A higher reading on one winding = possible open or corroded connection.
- A lower reading = shorted turns.

5. Continuity to Earth

- Test each terminal to motor frame
- Should be open circuit

6. Rotation Check

- Apply short pulse of 3-phase power (if safe to do so).
- Observe rotation direction.
- Swap any two phases to reverse if needed.

Real-World Fault Example: Pump Motor Trips Breaker on Start

Scenario:
A 5.5 kW pump motor tripped the circuit breaker immediately every time it was started. No motor movement or sound was observed, the breaker tripped the moment the start button was pressed.

Initial Checks:

- Verified 400V supply present on all three phases at the motor isolator.
- Confirmed breaker size and cable rating were appropriate for motor load.

Testing:
Removed the terminal links and performed insulation resistance (IR) test between each phase and earth:

- U–Earth: 0.1 MΩ
- V–Earth and W–Earth: similar low readings

This reading was far below acceptable limits (typically >1 MΩ minimum), indicating insulation failure or moisture ingress.

Investigation:
Opened the motor terminal box for a visual inspection. Found clear signs of water ingress, moisture and corrosion around the terminals.

Retest After Drying:
Tried drying the motor and re-testing insulation resistance. After drying:

- IR readings still below 1 MΩ, not safe for energising.

Fix:

- Motor replaced with a new unit.
- The replacement motor was insulation resistance tested before installation.
- Faulty motor sent for professional rewind and refurbishment.

Outcome:
New motor installed and commissioned without issue. Protective device held on start, and the pump ran under normal load. The failed motor was confirmed to have internal moisture damage beyond field repair.

Key Lesson:
Low insulation resistance can result in immediate operation of protective devices when a motor is energised. Always IR test suspect motors before energising, especially after exposure to moisture or poor storage conditions. Even if the motor appears dry, internal winding damage may require a full rewind or replacement.

Pro Tips:
Check the motor terminal box carefully. Loose terminal links, damaged lugs, corrosion, and poor terminations are common causes of motor faults and intermittent operation.

Many electrical faults generate heat long before they fail completely. Use a thermal imaging camera to identify hotspots early and rectify issues before they arise.

Single-Phase Motor Fault Finding

Overview
Single-phase motors are commonly found in domestic and light commercial equipment, such as fans, pumps, air conditioners, and small machinery. Unlike three-phase motors, single-phase motors require a start mechanism (capacitor or auxiliary winding) to create a rotating magnetic field.

Most faults can be grouped into:

1. Electrical faults
2. Mechanical faults
3. Control/start circuit faults

Accurate testing and diagnosis save unnecessary motor replacements and downtime.

Motor Testing Procedure

1. Visual Inspection

- Check the motor nameplate for correct voltage and wiring diagram.
- Look for burn marks, swollen or leaking capacitors, cracked insulation, or signs of overheating.
- Check cooling vents for blockages and dirt buildup.

2. Supply Check

- Verify that start/stop controls are working properly.
- For capacitor-start motors, inspect the capacitor visually, look for bulges or leaks.
- If fitted, inspect the centrifugal switch or electronic start relay for signs of damage or sticking, as these are common causes of starting faults.

3. Insulation Resistance Test (Megger Test)

Use a 500V insulation tester.

- Test each winding to earth and winding-to-winding.
- Typical healthy reading: >100 MΩ.
- If <1 MΩ, the winding may be damaged or contaminated.

4. Winding Resistance Test

- Use a low-ohm meter or multimeter.
- Test main and start windings separately:
- Main winding should have lower resistance.
- Start winding typically has higher resistance due to thinner wire.
- Compare with manufacturer specs if available.
- Open winding → infinite resistance
- Shorted winding → unusually low resistance

5. Capacitor Test

- Use a capacitance meter to check if the capacitor value matches the nameplate or motor specs.
- A failed capacitor is a common reason a single-phase motor won't start or hums but doesn't turn.

Pro Tip:

Always discharge capacitors before testing. Use a suitable resistor or follow the manufacturer's recommended discharge procedure before handling or measuring the capacitor.

6. Mechanical Check

- Rotate the shaft by hand, it should spin freely without binding or excessive noise.
- Listen for bearing noise or signs of rotor rubbing.

Real-World Fault Example: Fan Motor Hums but Won't Start

Scenario:
A small exhaust fan motor would hum loudly but never spin up to speed.

Initial Checks:

- Verified correct supply voltage at motor terminals.
- Inspected capacitor, found bulging and oil leak.

Testing:

- Measured capacitance: 0 μF (open circuit) instead of the rated 6 μF.
- Checked windings for continuity: both start and run windings OK.

Fix:

- Replaced the faulty capacitor with a new one of correct rating.

Outcome:

- Fan started immediately and ran at normal speed.
- The motor completed several start-stop cycles without further issues.

Key Lesson:
A failed start or run capacitor is one of the most common causes of a single-phase motor humming without starting. Always test the capacitor before condemning the motor or its windings.

6

CONCLUSION AND FINAL CHECKS

Verifying Repairs and Ensuring Safe Operation

Completing a repair is only half the job, verifying that the system is safe, reliable, and compliant is just as important. This section ensures you leave the job site knowing the issue is truly fixed and nothing else was introduced in the process.

Final Verification Checklist

- Visual Re-inspection
  - No loose terminals, wires, or components.
  - Cable insulation intact and properly dressed.
  - Correct tools removed from enclosures.
  - No foreign objects or debris left in distribution boards.
  - All covers, barriers, and enclosures securely refitted.
- Electrical Testing
  - Earth continuity verified.
  - Insulation resistance checked (where applicable).
  - Confirm correct voltages present on all terminals.
  - Check earth fault loop impedance, RCD operation, or other tests where required.
  - No signs of phase imbalance.
- Functional Testing
  - Operate the circuit/system under normal load.
  - Monitor start-up and shut-down behaviour.
  - Check any auxiliary functions (indicator lights, buzzers, fault indicators, etc).
  - Monitor for abnormal noise, vibration, overheating, arcing, or unexpected operation.

- System-Specific Tests
  - HVAC: Confirm temperature control and fan speeds cycle correctly.
  - Motors: Monitor current draw, noise, vibration, and temperature.
  - PLC Panels: Confirm correct input/output activity and sequence.
  - Appliances: Test full functionality, including safety features.

- Labeling & Documentation
  - Update any fault tags, isolation labels, or lock off records.
  - Mark replaced parts clearly if needed.
  - Record test results in logbooks or maintenance systems.

- Client/operator communications
  - Explain what was wrong and how it was fixed.
  - Advise if parts need follow-up maintenance or attention.
  - If applicable, demonstrate how to safely operate or reset equipment.

Pro Tip:

Walk Away With Confidence.

If you're not comfortable leaving the site without watching the system run through at least one full cycle, then you're not finished yet. **Remember it is your license; have you left everything to the regulatory standard?**

Maintenance and Preventative Measures

Prevention is always better (and cheaper) than cure. Regular maintenance not only reduces breakdowns and costly downtime but also helps detect small issues before they become major faults. This section covers essential preventative practices for electrical systems across commercial and industrial settings.

General Preventative Measures

1. Routine Visual Inspections
   - Look for signs of heat damage, corrosion, or wear.
   - Check tightness of terminals and signs of arcing.
   - Ensure enclosures are sealed, especially outdoors.
2. Thermal Imaging
   - Use infrared cameras to detect hot spots in switchboards, motor terminals, and control panels.
   - A thermal scan can reveal loose connections or overloaded components before they fail.
3. Torque Checks
   - Schedule torque checks on busbars, contactors, and cable lugs annually or as per manufacturer specifications.
   - Vibration, load cycling, and thermal expansion all loosen terminals over time.
4. Filter Cleaning & Cooling Checks
   - In HVAC systems and control panels, dirty filters and blocked cooling fans are a major cause of overheating.
   - Keep ventilation paths clear to prolong component life.
5. Test RCDs and Breakers Regularly
   - RCDs should trip within the time specified by BS 7671:2018+A4:2026.
   - Also consider insulation and earth fault loop impedance testing during routine maintenance.

7

GLOSSARY AND QUICK REFERENCES

Glossary

AC (Alternating Current) – Electrical current that reverses direction periodically, typically used in power grids.

Appliance – A device that uses electricity to perform a task, such as a heater or oven.

Arc Flash – A sudden release of energy caused by an electrical fault, creating intense heat and light.

Backfeed – When electricity flows in the reverse direction of the intended supply path, usually from an alternate source (like a generator, solar inverter, or UPS) back into the system.

Capacitor – A component that stores and releases electrical energy in a circuit.

Circuit – A complete path for electrical current to flow from source to load and back.

Circuit Breaker – An automatic device that interrupts power when it detects an overload or short circuit.

Client – The person or organisation requesting or receiving electrical services.

Coil – A wire wound into a spiral, often used in relays, transformers, and motors.

Commercial Installation – Electrical work involving shops, offices, and businesses.

Compressor – A device used to increase pressure in HVAC and refrigeration systems, typically motor-driven.

Contactor – An electrically operated switch used for switching a power circuit, especially in motors.

Control Board – A circuit board that manages functions in appliances or machinery.

Current – The flow of electrical charge through a conductor, measured in amperes (A).

Delta – A method of connecting three-phase motor windings in a delta (Δ) configuration, where the windings are connected in a closed loop.

Diagnose – The process of identifying the cause of a fault or problem in electrical equipment.

Distribution Board – A panel housing fuses, switches, breakers, and controls for power distribution.

DOL (Direct-On-Line) – A motor starting method where full voltage is applied directly to the motor terminals.

Domestic Installation – Electrical work in residential homes or small dwellings.

Earth – A safety path for electrical current to return to the earth and then back to the neutral point in case of a fault.

Element – A resistive component in appliances that converts electricity into heat.

Electricity – The flow of electrical energy created by the movement of electrons.

Emergency Stop (E-Stop) – A switch that immediately cuts power to machinery in case of emergency.

Fuse – A safety device that melts and breaks the circuit when current exceeds a set value.

Hydraulic – A system that uses pressurised fluid to perform mechanical work.

HVAC – Heating, ventilation, and air conditioning systems.

Impedance – The total opposition a circuit offers to AC, including resistance and reactance.

Industrial Installation – Electrical work in factories, large-scale plants, or heavy machinery settings.

Insulation – Material that prevents the unintended flow of current from a conductor.

Isolation – The process of separating a circuit or equipment from its power source for safety.

Junction Box – An enclosure that protects and joins electrical conductors.

Line Conductor – A wire that carries current from the power source to a device or circuit.
Locking Off – A safety procedure ensuring power is disconnected and cannot be reconnected during maintenance.
MCB (Miniature Circuit Breaker) – A compact device that protects circuits from overload and short circuit.
MCC (Motor Control Centre) – A centralised room housing motor starters, protection devices, and control gear.
Motor – A device that converts electrical energy into mechanical motion.
Multimeter – A tool used to measure voltage, current, and resistance in a circuit.
Overload – A condition where too much current flows through a device or conductor.
Pneumatic – A system that uses compressed air to transmit energy or do work.
PLC (Programmable Logic Controller) – A digital device used to control machines and processes.
PPE (Personal Protective Equipment) – Safety gear such as gloves, goggles, and helmets used to reduce injury risk.
Phase – One of the alternating current conductors in an AC supply. Three-phase systems are commonly used in commercial and industrial installations.
Power Surge – A sudden increase in voltage that can damage equipment.
Pump – A mechanical device that moves fluids, often powered by an electric motor.
RCD (Residual Current Device) – A safety device that disconnects power when it detects leakage current to earth.
Relay – An electrically operated switch used for control or isolation.
Resistance – The opposition to current flow in a material, measured in ohms (Ω).
Rotor – The rotating part of a motor or generator.

Sensor – A device that detects changes in environment (e.g., temperature, pressure) and sends signals.
Short Circuit – A fault where current flows through an unintended, low-resistance path.
Star – A method of connecting three-phase motor windings where one end of each winding is connected together to form a common neutral point.
Temperature – A physical quantity expressing heat level, often monitored in fault diagnostics.
Testing – The process of measuring or checking electrical systems for faults or compliance.
Thermostat – A device that senses temperature and controls heating or cooling systems.
Transformer – A device that changes the voltage level of an electrical supply.
Upstream – Refers to components or circuits closer to the power source.
Voltage – The electrical potential difference between two points, measured in volts (V).
VSD (Variable Speed Drive) – A device that controls motor speed and torque by varying input frequency and voltage.
Windings – Coiled wires in motors, generators, or transformers that create magnetic fields.

References and Standards

The following references and standards were used throughout the development of this book to ensure alignment with best practices, regulatory compliance, and industry expectations across the United Kingdom.

- BS 7671:2018+A4:2026 – Requirements for Electrical Installations (IET Wiring Regulations)
  - Core reference for electrical installation design, protection, inspection, testing, verification, isolation, and fault-finding practices within the UK.
- Electricity at Work Regulations 1989 (EAWR)
  - Primary UK legislation governing safe working on or near electrical systems, including duties relating to live work, competence, and safe systems of work.
- HSE HSG85 – Electricity at Work: Safe Working Practices
  - Health and Safety Executive guidance on planning electrical work, working dead versus live, isolation procedures, PPE, and risk control measures.
- HSE GS38 – Electrical Test Equipment for Use on Low Voltage Electrical Systems
  - Guidance on the safe selection and use of electrical test instruments, including probes, leads, and fused test equipment.
- IEC 61010-1 – Safety Requirements for Electrical Equipment for Measurement, Control, and Laboratory Use
- Defines safety requirements and measurement categories (CAT ratings) for electrical test equipment.
- Electrical Safety First – Best Practice Guide 2: Safe Isolation for Low Voltage Installations.

References and Standards

- IEC 60479-1 – Effects of Current on Human Beings and Livestock
  - Source reference for the physiological effects of electric current, including shock severity and time/current relationships.
- IET Code of Practice for In-service Inspection and Testing of Electrical Equipment
  - Industry guidance commonly used in the UK for the inspection and testing of portable and workplace electrical equipment (PAT testing).
- BS EN 60204-1:2018+A1:2025 – Safety of machinery. Electrical equipment of machines.
- BS 5266-1 – Emergency and Exit Lighting
- BS EN 61439 – Low-voltage switchgear and controlgear

Diagram Symbols

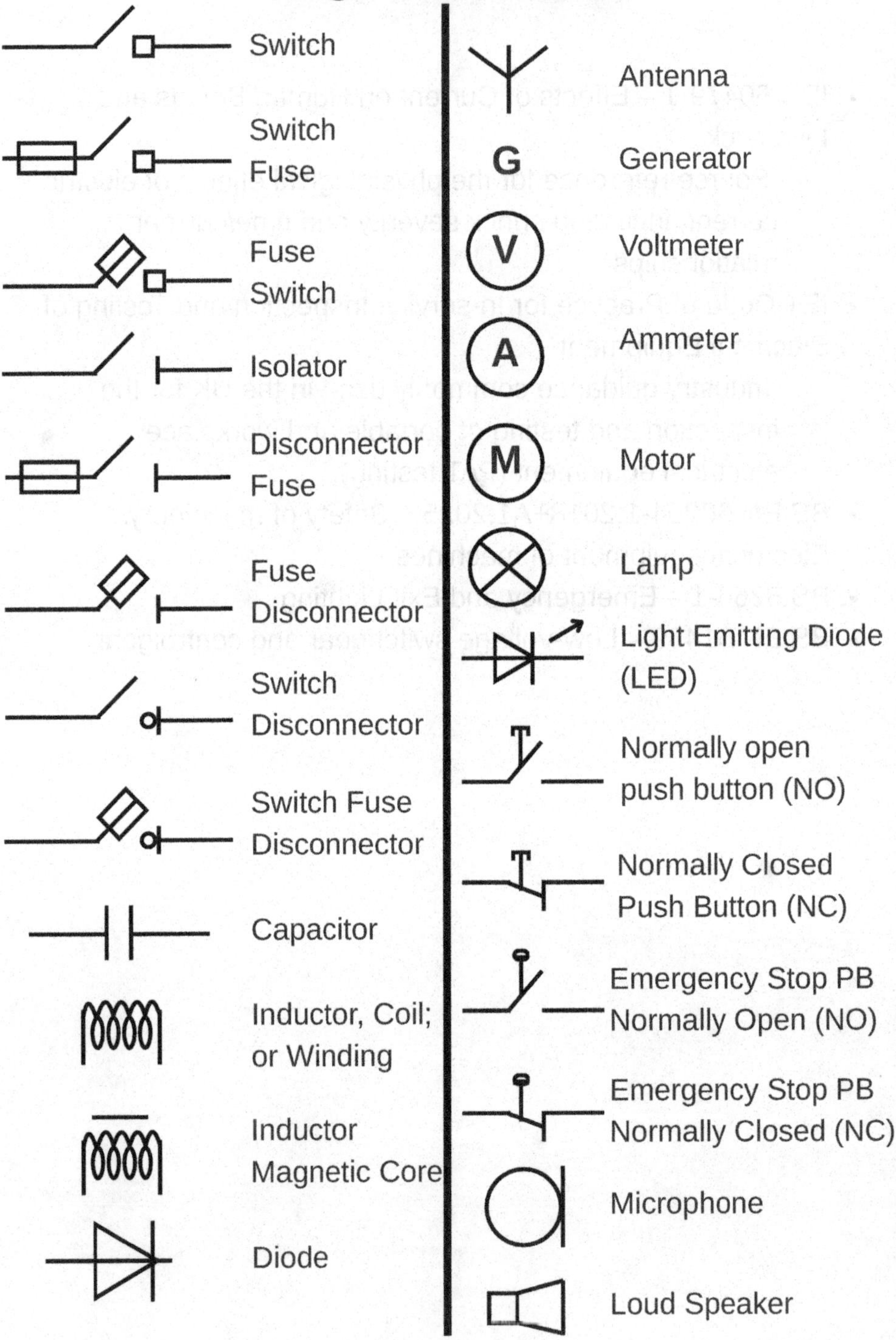

Diagram Symbols

Pull Switch (NO)

Pull Switch (NC)

Rotary Switch (NO)

Rotary Switch (NC)

Transistor (PNP)

Transistor (NPN)

Plug - Male

Coax Plug - Male

Socket - Female

Wire Connection

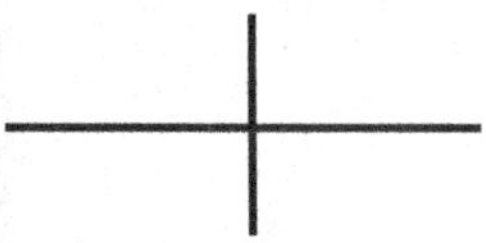
Wires Crossing (Not Connected)

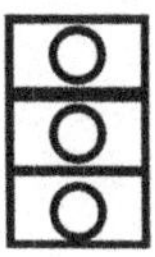
Terminal Block

Circuit Breaker

Limit Switch (NO)

Limit Switch (NC)

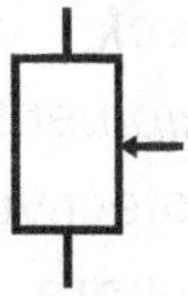
Potentiometer

Earth Connection

Protective Earth

Chassis Earth

Equipotentiality

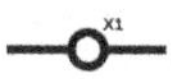

Terminal Connector

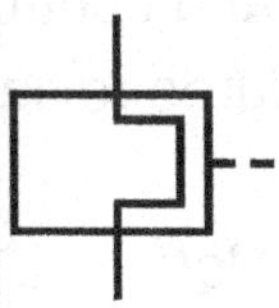
Thermal Overcurrent

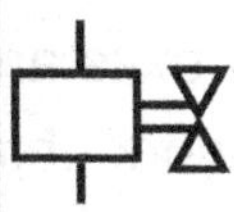
Solenoid Valve

Electrical Safety Checklist

Use this checklist before, during, and after electrical work to ensure safety compliance.

1. Pre-Work Safety Check

Personal Protective Equipment (PPE) Ready:

- Safety boots with protective toe caps.
- Insulating gloves where required for the task and risk assessment.
- Safety glasses/face shield for eye protection.
- Fire-resistant clothing (especially for arc flash work).
- Hearing protection (earplugs or earmuffs if in loud environments).
- Dust mask/respirator (if working in dusty areas).
- Hard hat (if required for overhead hazards).

Tools & Equipment Checked:

- Voltage tester/multimeter functional and calibrated.
- Insulated tools available where required.
- Safe isolation kit available (lock-off device, warning tag, proving unit, and approved voltage indicator).

Risk Assessment Completed:

- Hazards identified and control measures in place.
- Emergency procedures reviewed.

Sign-In Requirements Met:

- Signed in on-site.
- Aware of emergency exits and muster points.
- Relevant permits to work obtained.

2. Work Execution Safety Check

Safe Isolation:

- Power source identified and switched off.
- Lockout device and warning tag applied.
- Tested for dead before starting work.

Safe Work Practices:

- Never touching live wires or exposed conductors.
- Avoiding wet conditions or standing water.

Work Area Safety:

- Work area is well-lit and organised.
- Cables properly managed to avoid tripping hazards.
- Ladders secured and fibreglass if near live conductors.

3. Post-Work Safety Check

Re-Energisation Procedure:

- All tools and personnel clear of circuits.
- Lock-off devicesand warning tags removed only by the person who applied it.
- Equipment inspected and confirmed safe before re-energisation.
- Covers, barriers and guards refitted.

Final Worksite Inspection:

- Area cleaned and free of hazards.
- Equipment properly stored.
- All documentation and test results completed.
- Signed out of site.

Incident Reporting:

- Any near misses or hazards reported.
- Injuries recorded and reported as per site policy.

STEP	**ACTION**	**PURPOSE**
1. Initial Assessment	Identify the symptoms of the issue by gathering information from reports, users or observation.	Understand the nature of the fault and narrow down potential causes.
2. Safety Checks	Isolate power, check for hazards, wear appropriate PPE, follow Isolation procedures, verify the absence of voltage, ensure proper earthing, assess risks.	Prevent injury and ensure compliance with safety protocols.
3. Diagram Analysis	Refer to electrical schematics and wiring diagrams to understand how the system operates.	Identify possible failure points based on circuit design.
4. Testing	Use a multimeter, insulation tester or clamp meter to do basic checks on voltage, continuity, resistance and current.	Gather measurable data to locate abnormalities.
5. Diagnosis	Analyse test results, compare expected and actual readings, and eliminate unlikely causes.	Isolate the fault and determine the root cause.
6. Repair	Replace or fix faulty components, reconnect loose wiring, or reset tripped breakers.	Restore system functionality.
Final Checks	Reapply power, test under normal operating conditions, and monitor performance.	Ensure the issue is fully resolved and prevent recurrence.

Fault Finding Checklist

1. Initial Assessment – Identify the Problem

- What exactly isn't working? (Write below)

- When did the fault start? (Write below)

- Is it constant or intermittent? (Write below)

- Has anything changed recently? (e.g. new installs, weather, renovations) (Write below)

- Take photos as you go for identification, etc.

2. Safety Checks

- Lockout/Tagout completed (if needed)? ☐
- Appropriate PPE worn? ☐
- Confirm circuit is de-energised before handling? ☐
- Environmental risks identified? (e.g. water, confined spaces) ☐

3. Diagram Analysis

- Locate and review circuit or equipment diagram ☐
- Understand how the system is meant to operate ☐
- Identify key components and connections involved ☐
- Highlight potential fault zones based on symptoms ☐

4. Testing

- Verify supply voltage is present ☐
- Test continuity of wiring, switches, and terminations ☐
- Measure insulation resistance (IR) if applicable (e.g. motor windings) ☐
- Test RCDs, fuses, breakers, and load connections ☐
- Use loop/impedance testing if relevant ☐

5. Diagnosis

- Cross-check test results with diagram and symptoms ☐
- Rule out working components step-by-step ☐
- Narrow down to the likely cause ☐
- Confirm the fault before replacing parts ☐

6. Repair

- Repair, replace or isolate and lockout (with an out-of-service tag) the faulty component ☐
- Re-terminate connections securely ☐
- Remove temporary test links or bypasses ☐
- Restore full system function ☐

7. Final Check

- Power on system and verify correct operation ☐
- Retest all safety systems (e.g. RCDs, overloads) ☐
- Confirm circuit matches original design intent ☐
- Clean up and document your work ☐

NOTES:

NOTES:

Made in the USA
Monee, IL
13 September 2026

62981952R00079